Carbon Dots as Theranostic Agents

Scrivener Publishing
100 Cummings Center, Suite 541J
Beverly, MA 01915-6106

Advances in Nanotechnology & Applications

Series Editor: Madhuri Sharon

The unique properties of nanomaterials encourage the belief that they can be applied in a wide range of fields, from medical applications to electronics, environmental sciences, information and communication, heavy industries like aerospace, refineries, automobile, consumer and sports good, etc.

This book series will focus on the properties and related applications of nanomaterials so as to have a clear fundamental picture as to why nanoparticles are being tried instead of traditional methods. Since nanotechnology is encompassing various fields of science, each book will focus on one topic and will detail the basics to advanced science for the benefit of all levels of researchers.

Series Editor: Madhuri Sharon, Director, Walchand Centre for Research in Nanotechnology & Bionanotechnology
W.H. Marg, Ashok Chowk, Solapur 413 006
Maharashtra,
India

E-mail:sharonmadhuri@gmail.com

Publishers at Scrivener
Martin Scrivener (martin@scrivenerpublishing.com)
Phillip Carmical (pcarmical@scrivenerpublishing.com)

Carbon Dots as Theranostic Agents

Madhuri Sharon and
Ashmi Mewada

Scrivener Publishing

WILEY

This edition first published 2018 by John Wiley & Sons, Inc., 111 River Street, Hoboken, NJ 07030, USA and Scrivener Publishing LLC, 100 Cummings Center, Suite 541J, Beverly, MA 01915, USA

For more information about Scrivener publications please visit www.scrivenerpublishing.com.

Wiley Global Headquarters
111 River Street, Hoboken, NJ 07030, USA

For details of our global editorial offices, customer services, and more information about Wiley products visit us at www.wiley.com.

Library of Congress Cataloging-in-Publication Data

ISBN 978-1-119-46010-7

Cover image: Pixabay.Com
Cover design by Russell Richardson

Set in size of 13pt and Minion Pro by Exeter Premedia Services Private Ltd., Chennai, India

10 9 8 7 6 5 4 3 2 1

Contents

Preface **xi**

1 Carbon Dots: Discovery, Synthesis and Characterization **1**

1.1 Background 1

1.2 Introduction to QD 2

1.2.1 What is Quantum Mechanics? 4

1.2.2 Quantum Confinement 7

1.2.3 Discovery and History of Carbon Dots 8

1.3 Carbon QD and Graphene QD 9

1.4 Various Methods of Synthesis of Carbon Dots 10

1.4.1 Electrochemical Methods 11

1.4.2 Combustion and Thermal Oxidation Method 13

1.4.3 Hydrothermal Oxidation Method 15

1.4.4 Solvothermal Method 18

1.4.5 Laser Ablation of Graphite 18

1.4.6 Pulsed Laser Irradiation of Carbon Source 20

1.4.7 Arc Discharge Method 20

1.4.8 Plasma Treatment 21

1.4.9 Opening of Fullerene Cage 22

1.4.10 Ultrasonication Method 22

1.4.11 Microwave-Assisted Method 23

1.4.12 Chemical Methods 26

1.4.13 Supported Synthetic Procedure 26

1.4.14 Biogenic Method 28

1.5 Characterization of Carbon Dots 31

1.5.1 Microscopic Methods 32

1.5.1.1 SEM and TEM Characterization 32

1.5.1.2 AFM and STM Characterization 34

1.5.2 Spectroscopic Methods 35

1.5.2.1 UV-Vis Spectroscopy and its Application for Band Gap Determination 37

1.5.2.2 Fluorescence Spectrometry 37

1.5.2.3 Fourier Transform Infrared (FTIR) Spectroscopy 38
1.5.2.4 X-Ray Diffraction (XRD) Analysis 40
1.5.2.5 X-Ray Photoelectron Spectroscopy (XPS) 41
1.5.2.6 Dynamic Light Scattering/Photon Correlation Spectroscopy (DLS/PCS) 41
1.5.2.7 Dual Polarization Interferometry (DPI) 42
1.5.2.8 Raman Spectroscopy 43
1.5.2.9 Nuclear Magnetic Resonance (NMR) Spectroscopy 44
1.6 Summary 45

2 Properties of Carbon Dots 47
2.1 Introduction 47
2.2 Optical Properties 49
2.2.1 Absorbance 51
2.2.2 Photo-Induced Electron Transfer (PET) with CDs 52
2.2.3 Fluorescence/Photoluminescence (PL) 53
2.2.3.1 Multiphoton Excitation 60
2.2.3.2 Upconversion Photoluminescence 61
2.2.3.3 Lack of Blinking 63
2.2.3.4 Resistance to Photobleaching 63
2.2.4 Photocatalytic Property 65
2.3 Chemically Inert 66
2.4 Easy Functionalization 67
2.5 Water Solubility 67
2.6 Low Toxicity 68
2.7 Biocompatibility 69
2.8 Summary 69

3 Carbon Dots and Conjugates 71
3.1 Introduction 71
3.2 Why Conjugation of Carbon Dots? 74
3.3 Types of Carbon Dot Conjugates and Their Applications 76
3.3.1 Biogenic Compounds Conjugated with Carbon Dots 77
3.3.1.1 CDs Conjugated with Proteins/Peptides 78
3.3.1.2 CD Conjugates of Amino: Carboxylic Acid Ratio 80
3.3.1.3 CDs Conjugated with DNA 80
3.3.1.4 CDs Conjugated with RNase and SiRNA 84
3.3.1.5 CDs Conjugated with Lipid 86
3.3.1.6 CDs Conjugated with Folic Acid 86

3.3.1.7 CDs Conjugated with Chitosan 88
3.3.1.8 CDs Conjugated with Digitonin 89
3.3.2 Inorganic Heteroatoms Conjugated with CDs 90
3.3.2.1 CDs Conjugated with Gold Nanoparticles 91
3.3.2.2 CDs Conjugated with Silica 92
3.3.2.3 CDs Conjugated with ZnO 94
3.3.2.4 CDs Conjugated with CdS 95
3.3.2.5 CDs Conjugated with Strontium Oxide 96
3.3.2.6 CDs Conjugated with Gadolinium(III) 97
3.3.2.7 CDs Conjugated with Europium 97
3.3.2.8 CDs Conjugated/Doped with Nitrogen, Sulphur, Phosphorus and Boron 99
3.3.3 Carbon Dots Conjugated with Organic Material 100
3.3.3.1 PEG (Polyethylene Glycol) 101
3.3.3.2 CDs Conjugated with PEI (Polyethylenimin) or Polyaziridine 102
3.3.3.3 CDs Conjugated with α-Cyclodextrin 105
3.3.3.4 CDs Conjugated with Cysteamine 106
3.3.3.5 CDs Conjugated with Dihydrolipoic Acid 106
3.3.3.6 CDs Conjugated with Polyamidoamine (PAMAM) Dendrimers 107
3.3.3.7 CDs Covalently Conjugated with Rhodamine B Dyes 108
3.3.3.8 CDs Conjugated with Fe⁻ Aminoclay (FeAC) 109
3.3.3.9 CDs Conjugated with MWCNT 109
3.3.4 CDs Conjugated with Antibiotics 110
3.3.4.1 CDs Conjugated with Ciprofloxacin 111
3.3.4.2 CDs Conjugated with Tetracycline 114
3.3.4.3 CDs Conjugated with Vancomycin 114
3.3.4.4 CDs Conjugated with Ampicillin 115
3.3.4.5 CDs Conjugated with Streptomycin 116
3.3.5 CDs Conjugated with Anti-Neurodegenerative Drugs for Delivery to Central Nervous System 118
3.3.5.1 CDs Conjugated with Haloperidol 119
3.3.5.2 CDs Conjugated with Transferrin 124
3.3.5.3 CDs Conjugated with Curcumin 125
3.3.6 CDs Conjugated with Anticancer Drugs 128
3.3.6.1 CDs Conjugated with Doxorubicin 128
3.3.6.2 CDs Conjugated with Cisplatin 130
3.4 Summary 132

4 **CD as Drug Delivery Vehicle** **133**
4.1 Introduction 133
4.2 Considerations in Using CD as Drug Delivery Vehicle 136
4.3 Designs of CD-Based Drug Delivery System 137
4.3.1 Designing for Water-Insoluble Drugs 138
4.3.2 Designing for Targeting Tumor Location 138
4.3.3 Designing a Theranostic Nanomedicine 139
4.3.4 Designing a Photoresponsive Nano Drug Delivery System 139
4.3.5 Designing for Gene Delivery 140
4.3.6 Designing for Antibiotics Delivery 141
4.4 Carbon Dots for Delivery of Anti-Cancer Drug 142
4.4.1 A Brief Introduction to Cancer 143
4.4.2 Necessity of Drug Targeting in Cancer Therapy 144
4.4.3 Targeting Angiogenesis with CD 144
4.4.4 Various CD Conjugates for Delivering Anti-Cancer Drug 145
4.4.5 CD for pH-Dependent Drug Release 146
4.4.6 CD for Drug Delivery to Renal Cancer 147
4.4.7 CD for Drug Delivery to Lung Cancer 148
4.4.8 CD for Drug Delivery to Breast Cancer 149
4.5 CD for Drug Delivery to Neurodegenerative Disease 150
4.6 CD for Gene Therapy 151
4.7 CD to Monitor Delivery of SiRNA 152
4.8 Challenges in Using CD as Drug Delivery Vehicle 152
4.8.1 Prevention of Drug from Biological Degradation 153
4.8.2 Effective Targeting 154
4.8.3 Patient Compliance 155
4.8.4 Cost Effectiveness 155
4.9 Suitability of CD-Conjugated Drugs 156
4.9.1 For Oral Drug Delivery 156
4.9.2 By Inhalation 156
4.9.3 As Transdermal Drug Delivery 157
4.9.4 As Injection 157
4.10 Summary 157

5 **Carbon Dots for Cell Imaging and Diagnostics** **159**
5.1 Introduction 159
5.2 Bioimaging 162
5.2.1 Bioimaging of Cancerous Cells 166
5.2.1.1 HeLa Cells 168

5.2.1.2 Human Breast Cancer MCF-7 Cells and Human Breast Tumor Cells MDA-MB-468 170
5.2.1.3 B16F11 and HEK293 Cells 171
5.2.1.4 Ehrlich Ascites Carcinoma (EAC) Cells 173
5.2.1.5 Human U87 Cell 173
5.2.1.6 MGC-803 Human Gastric Cancer Cells 174
5.2.1.7 A549 Adenocarcinomic Human Alveolar (Lung) Basal Epithelial Cells 175
5.2.1.8 Human Hepatocellular Carcinoma Cells 175
5.2.1.9 Kidney Proximal Tubule Cell Line (LLC-PK1) 176
5.2.1.10 C6 Glioma Cells 177
5.2.2 Bioimaging of Nucleus 178
5.2.3 Bioimaging of Virus 180
5.2.4 Bioimaging of Bacteria 181
5.2.5 Bioimaging of *Drosophila melanogaster* 183
5.3 CDs as Sensor 184
5.3.1 Intracellular Detection of Ions 185
5.3.1.1 Detection of Ag Ions 185
5.3.1.2 CD for Detection of Cu Ion 187
5.3.1.3 Detection of Fe^{3+} Ions 190
5.3.1.4 Detection of Hg^{2+} Ions 191
5.3.2 Detection of Small Molecules 192
5.3.2.1 Detection of Nitric Oxide 193
5.3.2.2 Detection of Phosphate 193
5.3.2.3 Detection of Reactive Oxygen Species 194
5.3.2.4 Detection of H_2S 194
5.3.2.5 Detection of TNT 195
5.3.2.6 Detection of Hydroquinone 195
5.3.2.7 Detection of Surfactant 196
5.3.2.8 Detection of Humidity 197
5.3.3 Detection of Biological pH Value 198
5.3.4 Detection of Nucleic Acid 200
5.3.5 Detection of Vitamins 201
5.3.6 Detection of Protein and Enzymes 202
5.3.7 Detection of Glucose 202
5.3.8 Detection of Cancerous Cells 203
5.3.9 Detection of Dopamine Neurotransmitter 204
5.4 Concluding Remarks 206

6 Suitability of Carbon Dots as Payload for Plants **209**

6.1 Introduction 209

6.2 Suitability of Carbon Dots as a Payload for Plant 211

6.2.1 Cytotoxicity of Carbon Dots 212

6.2.2 Carbon Dots and Plant Growth 213

6.2.2.1 Wheat 215

6.2.2.2 Green Beans 215

6.2.3 Is Cell Wall a Barrier for Carbon Dot Internalization? 216

6.2.4 Other Possible Routes for Entry of Carbon Dots 219

6.3 Carbon Dots and Plant Fertilizer 221

6.3.1 Nitrogen 222

6.3.2 Phosphorus 224

6.3.3 Potassium 225

6.3.4 Micronutrients 226

6.4 Need for Sensor to Detect 227

6.4.1 Plant Nutrient Status 228

6.4.2 Water 229

6.4.3 Pathological Status of Plant 231

6.4.4 Residual Pesticide, Herbicide or Insecticide in Plants 232

6.5 Carbon Dots and Foliar Application of Drugs on Plants 233

6.5.1 Cuticular Layer 233

6.5.2 Cell Wall 235

6.5.3 Plasma Membrane 235

6.6 Carbon Dots as Trojan Horse to Penetrate Foliar Surface for Therapeutic Molecule Delivery 236

6.6.1 CD for Delivery of Antibiotics to Plants 237

6.7 Concluding Remarks 240

References **241**

Index **265**

Preface

I would like to begin this preface with a quote from Sir Isaac Newton, which justifies our efforts in writing this book i.e. **"What we know is a drop, what we don't know is an ocean"**. Carbon Dots (CD) is an emerging luminescent nanomaterial, to which scientists have employed tremendous efforts to study its use in health care systems, as a nontoxic alternative to replace traditional heavy metal-based Quantum Dots (QD). In this book, we have tried to compile the knowledge available for basic understanding of quantum mechanics and concepts related to the phenomenon, the discovery, properties, applications, as well as the present drawbacks of QD. CDs have many advantages, including low toxicity, chemical inertness, biocompatibility and no photobleaching that makes it a suitable theranostic agent for humans. In addition, typical CDs contain a large amount of hydrophilic carboxyl and hydroxyl groups, imparting them with excellent water solubility and multifunctionality. For enhancing the efficacies of CD for various applications, during synthesis, CDs have been conjugated with inorganic, organic, biogenic and synthesized drugs and antibiotics. Moreover, efforts are devoted to the surface passivation of CD with different molecules, so as to make it stable and at the same time also help in increasing the quantum yield, which is one of the most important aspects for biological applications. We have also explored the possibility of use of CD as a payload for plants nutrients/fertilizers and therapeutic agents, especially through foliar application. The importance of the physical and chemical properties attributed to bioapplications (high photostability and chemical stability, low cytotoxicity and high quantum yield) is also touched upon in this book. Various other aspects such as characterization techniques, product characteristics, bioavailability, pharmacokinetics, stability, drug use, and toxicity in humans have also been systematically introduced.

However, CDs prepared from diverse methods exhibit large size distribution and photoluminescence non-uniformity, and the complex and time-consuming separation and purification severely limit their further applications, impeding further bio-applications of CDs. Moreover, the mechanism of photoluminescence is still unclear. This demands that an effective synthesis method with a high yield should be developed to fabricate CDs with a high quantum yield in a small size distribution, meaning that an understanding of CDs' photoluminescence phenomenon, especially their bright multiphoton emission, should be explored to facilitate their *in-vivo* applications.

We hope that this book will suffice the need of those who want to enter into the magnificent field of carbon dots for various applications in the health care system.

Madhuri Sharon
Ashmi Mewada
Mumbai
31st May 2018

1

Carbon Dots: Discovery, Synthesis and Characterization

The saddest aspect of life right now is that science gathers knowledge faster than society gathers wisdom.
Isaac Asimov

1.1 Background

Owing to never-ending human curiosity for search of answers to unsolved complex natural phenomena, Max Plank, a German scientist, first introduced the concept of discretization to explain black body radiation in a slightly different way from the already available (classical mechanics) explanation [1]. In the race to explain atoms, Rutherford's atomic model unintentionally implied that atoms could not exist because the electron should eventually fall into the nucleus as it loses energy in its orbital motion around the nucleus. Niels Bohr first used Plank's concept to explain an acceptable and stable atomic model [2, 3]. The origin of the new concept of scientific theory or quantum theory or quantum mechanics had begun, as this theory could explain various natural phenomena in an acceptable way. Quantum mechanics generally is used to explain atomic and subatomic interactions with the help of rigorous mathematical derivations formulated by the Austrian scientist Erwin Schrödinger [4].

Since the discovery of fullerene [5], carbon nanoscience has emerged. It has extended up to one-dimensional (1D) carbon nanotubes (CNTs) [6, 7] and most recently two-dimensional (2D) graphene [8]. The physics of these nanomaterials could be explained by quantum mechanics. It has been observed that there is a drastic change in optical and electronic properties of materials once they are in the order of a few nanometers as compared to the properties of bulk materials. This change in properties of nanomaterials is attributed to a term called "quantum confinement," i.e., when material becomes as small as the de Broglie wavelength of electron wave function [9–12]. This factor can be explained using rigorous mathematical derivations. The mathematical explanation coined by Schrödinger constituted quantum mechanics or quantum physics. In quantum mechanics all the experimentally measurable quantities of a particular system, generally expressed in terms of position and momentum, are replaced by operators which operate on the wave function of the system. For example, the Hamiltonian operator operating on a given wave function in quantum mechanics corresponds to total energy of a system similar to the Hamiltonian one of a given system in classical mechanics.

1.2 Introduction to QD

The term *QD* takes us to a difficult domain of Physics where a material follows wave theory concept that introduces us to very small particles. A fluorescent colloidal semiconductor nanocrystal was first discovered in the 1980s by Louis E. Brus of Columbia University and was later named as quantum dot (QD) by Mark Reed in 1988 [13] because of quantum confinement. Since then thousands of people have contributed to the synthesis, understanding, physics, chemistry and various applications of QD.

Generally, QDs fall in the size range of 1–10 nm. QDs are known for various optical and electronic properties like excitation of multiple fluorescence spectra, enhanced photostability, high quantum yield, photocatalysis, biological imaging, diagnostics and molecular histopathology [14–18]. Inorganic QDs of various types have been reported to date which include CdTe, CdSe, PbSe, GaAs, GaN, InP, and InAs [19–21]. However, major drawbacks in employing these QDs for biological applications are related to their toxicity as well as insolubility in water [22–24]; and also the tedious and expensive procedure of QDs synthesis. To overcome the above-mentioned drawbacks related to QDs, researchers are constantly trying their hands at various directions in pursuit of more compatible materials.

Do you know that we see these particles in everyday life when we sit in front of LED or LCD television? The TV screen has incredibly small crystals of 2–20 nm, called QDs. These QDs have a unique property that depending on their size they emit different colors, e.g., smaller QDs emit blue light and larger ones emit red light. When QDs are used in TVs they give better color accuracy. But their uses are not limited to TVs only. They have found applications in transistors, solar cells, LEDs, diode lasers and they can charge smart phones in 30 seconds, etc.

The shortest possible explanation of QDs is that they are semiconductor crystals having their *excitons* or electron and holes confined in all three dimensions of space. Consequently, QDs have electronic properties that are intermediate between those of bulk semiconductors and those of discrete molecules. Being very small in size, they possess larger band gap. This is because of the discreet energy level, which depends on the size and shape of the QD. Smaller QDs need more energy to confine excitons to a smaller volume and they spread out more, thus having higher band gap energy. These very small semiconductor particles can be compared to the Bohr radius of the excitons, i.e., separation of electron and holes. A QD is usually 1–10 nm in size that encompasses ~10–50 atoms in diameter. Structure-wise, QD includes a *Core*—the active part of QD—surrounded by a protective layer called the *Shell*, which is responsible for enhancing the stability and emission efficiency of the core, and *Surface ligands*, which are the outermost passivating, protective and chemically active layer.

Apart from QD there are a few other quantum confinement structures, i.e., *Quantum well* and *Quantum wire*, but the difference is that unlike QD, which has electron or hole confinement in three dimensions of the space, in the *Quantum wire* they are confined in two spatial dimensions and in *Quantum well* electrons or holes are confined in one direction that can freely propagate in one dimension.

1.2.1 What is Quantum Mechanics?

The QDs are nanometer scale particles that follow the quantum mechanical principle of confinement. To understand the quantum mechanics, let's look at the physics of QD (semiconductor). The band gap of QD varies from those of bulk material, conducting material and insulators.

What is band gap? The difference in energy between the conduction band and the valance band is called the *Energy gap* or *Band gap* Figure 1.1. The electrons can be excited across this band gap into the conduction band with an applied voltage or a photon or any other energy source. When the electron is excited from the valance band it leaves an empty space called a hole. The electron-hole pair is called an exciton. Once excited, the electron can move around in the conduction band, and the hole can move around in the valance band.

The energy bands are quantized at the atomic level, which is broadened into bands of energy. The levels inside the bands

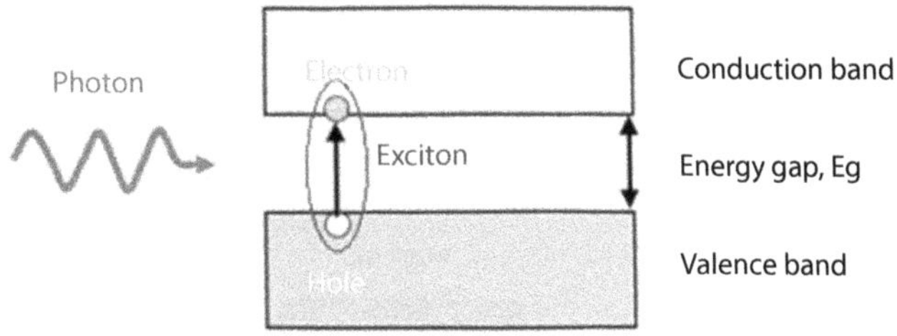

Figure 1.1 Band gap and exciton.

are very closely spaced. Hence, they are not considered to be discrete. Conducting materials conduct at room temperature because their band gaps are in the range of KT energy (~0.25 eV). Because of their low band gap, electrons from the valence band easily get excited to conduction band creating a corresponding number of holes in the valence band. These holes assist in making materials conductors even at room temperature.

In ***insulators***, though the valence band is full of electrons and the conduction band is completely empty of electrons, but both these bands are separated by a large energy (large band gap), which makes it difficult for electrons of the valance band to go to conduction band. A ***semiconductor*** is like an insulator, but the only difference is that the two energy bands (i.e., band gap) are in the vicinity of 1.0 eV to 3.0 eV, therefore, it is easier to excite an electron into the conduction band where it can move around. A schematic diagram of electron occupancy and allowed energy bands for given materials is shown in Figure 1.2.

Due to the small size of the QD, the energy bands narrow to discrete energy levels. The exciton is confined to the QD like a particle is in a box. The lowest energy level of the QD is given by the following equation:

$$E = E_{hole} + E_{electron} + E_g = \frac{\hbar^2\pi^2}{2m_h R^2} + \frac{\hbar^2\pi^2}{2m_e R^2} + E_g \quad (1.1)$$

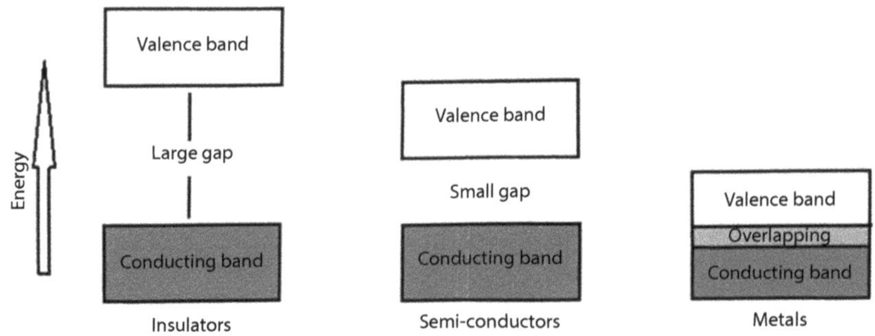

Figure 1.2 Schematic diagram of electron occupancy of allowed energy bands for given materials.

where E is energy; E_g is the semiconductor band gap; m_h is the effective mass of the hole, m_e is the effective mass of the electron, and R is the radius of the QD.

The equation for the energy of the hole and of the electron is the same as the particle in the box. The term E_g is present because the particles are not in an empty box; they are in a semiconductor so they have a zero-point energy. The smaller the size of QD, the larger will be the band gap.

de Broglie wavelength: The concept of matter waves in quantum mechanics reflects the wave-particle duality of QD. This theory was proposed in 1924 by Louis de Broglie. According to this theory, de Broglie relations show that the de Broglie wavelength is inversely proportional to the momentum of a particle; and the frequency of matter waves, as deduced by de Broglie, is directly proportional to the particle's total energy, i.e., the sum of particle's kinetic energy and rest energy.

Planck-Einstein equation: In 1900 Max Planck discovered the Planck constant, which is a physical constant reflecting the sizes of energy quanta in quantum mechanics. The Planck constant (h) is the proportionality constant between the energy (E) of a photon and the frequency (ν) of its associated electromagnetic wave. This relation between the energy and frequency is called the Planck-Einstein equation:

$$E = h\nu. \tag{1.2}$$

Since the frequency v, wavelength λ and speed of light $\boldsymbol{c}$ are related by $\lambda v = c$, the Planck relation is also expressed as

$$E = \frac{hc}{\lambda}. \tag{1.3}$$

The amount of energy is inversely related to the wavelength of the light; the shorter the wavelength, the greater the energy of each photon of the light.

1.2.2 Quantum Confinement

***Bohr's exciton radius*:** Electronic transitions within the material and light-scattering effects determine the optical properties of that material. Due to Coulomb interaction, the electrons and holes existing in a material are known to form excitons. Therefore, to understand the optical nature of semiconductors the properties of the exciton should be investigated. *The distance between the electron and the hole within an exciton is called the Bohr radius of the exciton.* The Bohr radius of semiconductors is a few nanometers. When the length of a semiconductor is reduced to the same order as the exciton radius (few nanometers), quantum confinement effect occurs and the exciton properties get modified. The QDs having radii larger than the exciton Bohr radius are in the "weak confinement regime" and those having radii smaller than the exciton Bohr radius are said to be in the "strong confinement regime." Thus, if the size of the quantum dot is < 10 nm then the quantum confinement effect dominates, the electronic and optical properties changes and the fluorescent wavelength is determined by the size.

If the radius of the QD shrinks to Bohr's exciton radius the motion of the electrons and holes are strongly confined spatially to the dimension of the QD. Consequently, the excitonic transition energy and the band gap energy increases, which results in the blue shift of the emission of the quantum dot.

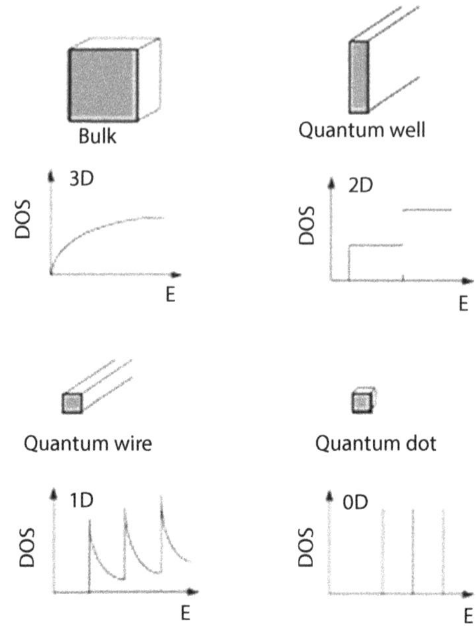

Figure 1.3 Density of states and their dimensions.

Spatial dimensions: Based on the dimension of the confinement, three kinds of confined structures are defined (Figure 1.3):

- Quantum Well (QW) – here the material size is reduced only in one direction and the exciton can move freely in the other two directions.
- Quantum Wire (QWR) – here the material size is reduced in two directions and the exciton can move freely in one direction only.
- Quantum Dot (QD) – here the material size is reduced in all directions and the exciton cannot move freely in any direction.

1.2.3 Discovery and History of Carbon Dots

The very first reported discovery of carbon dots (CDs) was the result of an accidental appearance of these fluorescent nanoparticles as a by-product during purification of single-walled carbon nanotubes (SWCNTs) by Xu *et al.* in 2004 [25]. This fluorescent fraction separated during purification was identified as a mere impurity in the crude nanotube soot which on further analysis displayed fluorescence in UV light. Other related analysis, such as Fourier transform infrared (FTIR) spectroscopy and energy-dispersive analysis of X-rays (EDAX), were performed by the researchers driven by their curiosity to understand these fluorescent particles (as it was termed during discovery) which showed the presence of carbon moieties on the surface and no metal residues from the catalysts used for synthesis of CNTs. Also, elemental analysis displayed highest carbon content followed by oxygen, hydrogen and nitrogen.

The discovery of CDs, which was a realization of a new form of fluorescent carbon nanostructure, attracted researchers to explore these newly found nanoparticles and understand their synthesis and properties. Later these carbon nanoparticles were termed as Carbon QD or Carbon Dots (CD) due to their striking similarities with traditional semiconductor QD like size, properties and photoactivity. In comparison with inorganic semiconducting QDs, CDs show enhanced properties of non-toxicity (biocompatibility), resistance to photo bleaching, water solubility and comparatively inexpensive synthesis protocol [26].

1.3 Carbon QD and Graphene QD

Carbon QD and Graphene QD both are known for optical properties like fluorescence and photoluminescence. However, these properties arise when an electron makes a transition from a higher electronic state to a lower one. This phenomenon is less frequent than thermal radiations, which are emitted by all the materials in a wide range of wavelengths. It is just because in bulk materials, excited electron has a lot of electronic states to descend into favoring release of heat or phonons. But when the same materials are physically confined the electronic states become less and are more widely separated from each other as compared to bulk, and this spatial confinement leaves electron no other choice than to de-excite with emission of photons. The word quantum is used with both carbon and graphene QD only because of this reason.

Graphene QD (GQDs) are nothing but small pieces of pure graphene sheets/flakes showing quantum confinement, thereby exhibiting fluorescence in a wavelength which is equal to the electronic band gap energy. One of the very first reports where graphene flakes were referred to as GQDs employed nanolithography technique for synthesis [27]. GQDs also contain various heteroatoms which play a major role in photoluminescence. On the other hand, CDs are mixture of sp^3 and sp^2 hybridized spherically shaped molecules of carbon and other heteroatoms. GQDs have followed mostly top-down methods for synthesis according to previous reports, whereas CDs are more inclined towards a bottom-up approach (mostly for obtaining best quality fluorescent nanodots). However, the cause of fluorescence in both nanodots is still debatable.

1.4 Various Methods of Synthesis of Carbon Dots

Ever since the discovery of CDs, a wide range of synthetic techniques have been developed for fabrication involving over a decade of research. These various approaches are broadly classified into two categories: “top-down” and “bottom-up” [28].

In the *top-down approach*, graphite, which is composed of many layers of graphene, is used as precursor. Graphite is delaminated into nanosized graphene via physical or chemical methods such as arc discharge, laser ablation, electrochemical oxidation and chemical oxidation, etc. These nano-size particles do not fluoresce. To produce light emission it is necessary to follow a passivation step, which is associated with defects in the graphene structures [29].

The *bottom-up approach* concentrates on synthesis of CDs from non-graphite carbon sources or molecular precursors (Lim *et al.* 2015)[30] by various chemical methods such as combustion, pyrolysis, microwave-assisted, ultrasonic, thermal, hydrothermal, electrochemical and biogenic methods [31]. Whereas, physical methods consist of arc discharge, laser ablation, plasma treatment and opening of fullerene cage, etc. Some of the main methods classified for synthesis of CDs are explained below.

1.4.1 Electrochemical Methods

Electrochemical procedure involves the use of a three-electrode cell containing working electrode, reference and counter electrode, as well as electrolyte and a carbon source. After application of a pre-decided potential and number of cycles, the resulting solution is subjected to purification to obtain CDs [32]. There are many different electrochemical methods used for synthesis of CDs.

Zhou *et al.* [32] fabricated CDs by oxidizing a graphitic column electrode against a saturated calomel electrode with a platinum (Pt) wire counter electrode in NaH_2PO_4 aqueous solution.

Using an assembly of a graphite rod as working electrode, a Pt mesh counter electrode, and an Ag/AgCl reference electrode immersed in pH 7 phosphate buffer solutions, Zheng *et. al.* [33] have synthesized CDs.

Bao *et al.* [34], using an electrochemical method, found enhanced control over synthesis of CDs as compared to other methods where the electrode potential and current density can be regulated. They explained a controllable process to obtain luminescent CDs by etching of carbon fibers. In a typical experiment, a conventional three-electrode cell was used with some carbon fibers as working electrode; a Pt wire and silicon (Si) wire were employed as counter and reference electrodes respectively. Acetonitrile solution containing 0.1M tetra-butyl-ammonium-perchlorate (TBAP) was used as electrolyte. The selected size of the CDs was obtained by merely regulating the applied potential at the electrode. The as-obtained CDs were purified by dialysis and no further surface passivation was required. The CDs displayed size- and excitation-dependent photoluminescence (PL) properties [34]. The main advantages of using electrochemical methods include cost effectiveness, simple operation, high output and also control over synthesis of CDs in terms of size and properties by regulation of electrode potential. However, more research and improvements are still required in terms of obtaining monodispersed and self-surface passivated CDs.

According to Que *et al.* [35] electrochemical oxidation of graphene electrode in phosphate buffer solution can also yield 3–5 nm uniform size CDs of 1–3 graphene layers due to strong inter-graphene attraction. The oxygen-containing groups on the surface of CDs enables aqueous solubility and facilitates surface functionalization.

Li *et al.* [36] tried to synthesize CD by ionic liquid (IL)-assisted electro-oxidation of graphite using the water-soluble IL 1-butyl-3-methylimi-dazolium tetrafluoroborate [BF_4] that has 90 wt% water as electrolytes.

An alkali-assisted electrochemical method is also used for generating 1–4 nm CDs [37]. In this method graphite rod is used as both anode and cathode, and NaOH/EtOH as electrolyte with a current intensity of 10–200 mA cm^{-2}. In this method graphite honeycomb layer is cut into ultra-small particles, leading to fragmentation of graphite into tiny high-quality CDs. It must be mentioned here that when sulphuric acid (H_2SO_4/EtOH) was used as electrolyte CDs did not form, thus confirming that alkaline environment and OH groups are essential for the formation of CDs by electrochemical oxidation process.

For large-scale synthesis of high-quality crystalline CDs with high purity, a method is proposed by Ming *et al.* [38] where pure water is used as an electrolyte without any other chemical additives. The CDs thus obtained show excellent aqueous dispersibility, and remarkable down- and up-converted PL properties; above all, it requires no further purification.

Yao *et al.* used pure graphite ring as anode and a titanium tube as cathode separated by an insulative O-ring. Pure water was used as electrolyte. Electrolytic voltage and ultrasonic power were applied at the same time to obtain blue fluorescent CDs, which did not require any further purification procedure. A quantum yield of 8.9% was observed. The CDs displayed excellent thermodynamic stability in aqueous solution [39]. They designed a sono-electrochemical method which yielded fluorescent CDs which were thermodynamically stable and highly fluorescent. Compared to normal electrolysis, high-intensity electric field in combination with ultrasonic field played a very important role in increasing the purity of CDs as well as enhancing the electrolytic efficiency.

1.4.2 Combustion and Thermal Oxidation Method

In one of the early reports, Liu *et al.* demonstrated synthesis of CDs by oxidative acid treatment of candle soot [40]. The combustion soot was obtained by placing a glass plate over a burning candle. The soot was then subjected to acidic treatment, which led to formation of hydrophilic CDs by decomposition of carbon aggregates into small size nanoparticles. The CDs were purified using polyacrylamide gel electrophoresis (PAGE), which in turn enhanced their luminescent properties.

Later, Tian *et al.* [41] used soot of combusting natural gas and paraffin oil as precursor to obtain highly fluorescent CDs. The soot was refluxed with nitric acid and the solution was subjected to centrifugation and prolonged dialysis to obtain purified CDs. By this method 4.8 ± 0.6 nm CDs were obtained. These CDs were very much similar to CDs obtained from candle soot. It is very important to notice one thing, which is that all the procedures which followed employment of carbon soot utilized acid oxidation method for procuring CDs (i.e., it was a two-step procedure of combustion followed by acid oxidation). Thermal decomposition methods can directly produce functionalized CDs but requires a suitable carbon source.

Simultaneously, Ray *et al.* [42] reported synthesis of CDs from carbon soot by acid oxidation using nitric acid. They also concentrated on purification and separation of CDs based on size by employing high-speed centrifugation.

Bourlinos *et al.* [43] have reported synthesis of hydrophilic CDs by thermal method. The authors have displayed synthesis of both hydrophilic and organophilic CDs using two different types of ammonia-containing citrate salts. A single-step thermal decomposition method was employed to obtain surface functionalized CDs below 10 nm, which showed excitation-dependent fluorescence spectra. In another report, chemical synthesis of CDs using citric acid as precursor is reported by the same research group. The surface functionalization was obtained by thermal method to get highly photoluminescent and hydrophilic CDs in the size range of 10–20 nm.

These methods are not fully effective because during thermal decomposition a high amount of insoluble carbonaceous materials are also produced, which in turn decrease the quantum yield. Also, the soot-based acid oxidation methods require separate gel-based purification procedures, which make it a lengthy, time-consuming process.

In another thermal oxidation process, CDs are synthesized by pyrolysis of ethylenediamine tetra acetic acid salts by controlled carbonization of sucrose (Wu *et al.* 2010)[43].

Chen *et al.* [44] attempted to synthesize CDs in a Mg-substituted microporous aluminophosphate (MAPO-44) molecular sieve having a chabazite structure through thermal decomposition of the occluded template or loaded organic molecules. Thus, resulting composite phosphors could be excited by a broad range of light in the ultraviolet region; the emission wavelength of this CD is tunable by varying the thermal treatment condition.

There are two more common thermal oxidation methods, namely, hydrothermal oxidation and solvothermal oxidation.

1.4.3 Hydrothermal Oxidation Method

This method has given clarity about the formation of CDs through four consecutive stages, i.e., dehydration, polymerization, carbonization, and passivation. Hsu and Chang [45] used four different molecules (glycine, cadaverine, 2-amino-2-hydroxymethyl-propane-1,3-diol (TRIS), and ethylene diamine-tetraacetic acid (EDTA)) containing either an amino group or a carboxyl group or both in aqueous solutions. Each was separately calcined hydrothermally at 300 °C for 2 h (Figure 1.4). The precursors possessing both amino and carboxyl groups were better for the preparation of highly water-dispersible and PL CDs. The quantum yield (QY) values of CDs synthesized from glycine, EDTA, TRIS, and cadaverine were 30.6%, 26.6%, 26.0%, and 5.4%, respectively.

This is one of the most commonly used methodologies for synthesis of CDs because of its simplicity and high yield. In this process several hydrocarbon precursors have been successfully used as precursor, for example:

- **L-ascorbic acid:** Zhang *et al.* [46] have reported synthesis of highly fluorescent very small sized

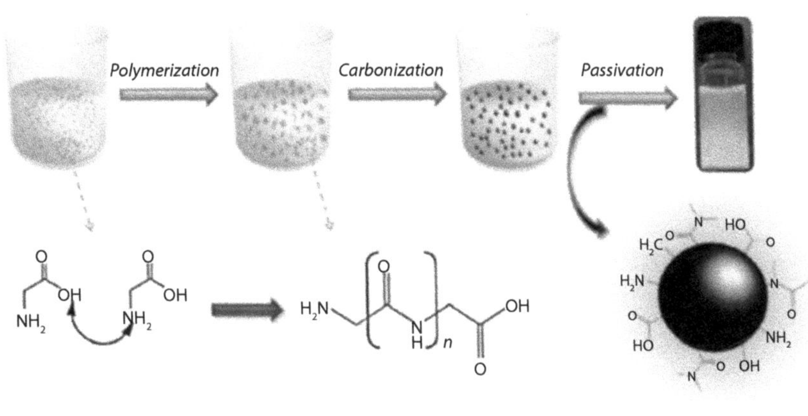

Figure 1.4 The four stages of CDs formation using glycine as a model precursor molecule: dehydration, polymerization, carbonization, and passivation. (Reprinted with permission from [45]; Copyright © 2010 John Wiley and Sons)

CDs using hydrothermal method. Aqueous solution of L-ascorbic acid was used as carbon source and ethanol was used as solvent to form a perfect reaction mixture to obtain uniformly shaped CDs at 90 °C. The CDs exhibited excitation-dependent fluorescence spectra. Interestingly, as compared to other methods no strong acid or alkaline treatment or surface modification was required for obtaining stable CDs. These CDs were found to possess excitation-, pH- and polarity-dependent luminescence, as well as upconversion fluorescence properties.

- **Carbohydrates:** A one-step synthesis of CDs via hydrothermal oxidation of carbohydrates like starch, glucose and sucrose in the presence of acid and alkali was reported [47]. The elemental analysis of the obtained CDs revealed carbon and oxygen dominance, which increased the water solubility property. The authors speculated that addition of acid (HCl) enhances emission at longer wavelengths whereas addition of a base (NaOH) enhances the emission at shorter wavelengths. Zhang *et al.* [48] used lactose as carbon source and Tris as a surface passivating agent. The solution was subjected to stirring at 100 degrees for 24 hours and the obtained CDs had amorphous carbonaceous structure. The CDs displayed hydroxyl and carboxyl groups, which were confirmed using FTIR.
- **Gelatin:** In a recent report, Wei *et al.* [49] have displayed synthesis of CDs using gelatin as precursor by simple hydrothermal method. The CDs were found to be stable and highly fluorescent. The obtained CDs were used as sensor to detect hydrogen peroxide and glucose based on their quenching properties.

- **Sorbitol:** Sharon's group [50] similarly reported the synthesis of CDs using sorbitol solution and NaOH-ethanol mixture. The CDs were further surface functionalized with bovine serum albumin (BSA) to render stability and enhance the fluorescent property. The BSA layer also acted as a linker for attachment of drug molecules and the whole conjugate was used for drug delivery experiments.
- **Proteins:** Apart from utilizing sugars as carbon source, proteins have also been experimented on for synthesis of CDs. For instance, Zhang *et al.* [51] have reported one-step synthesis of CDs using BSA by hydrothermal treatment.
- **Folic Acid:** In another report, folic acid was used to synthesize CDs by one-step hydrothermal method [52]. The aim of the study was to prepare functionalized CDs to target towards folate receptors, which are overexpressed in cancer cells. These CDs would therefore serve as targeting agent as well as therapeutic agent for cancer.
- **Carbon fibers (CF) and Carbon nanotubes (CNT):** *Hydrothermal cutting* of graphene sheets has also yielded 2–3 nm CDs. A facile one-step wet chemical synthesis of CDs by acidic oxidation of carbon fibers (CF) has been achieved [53]. Interestingly, the PL from the CF-derived CDs of different sizes can be tailored by simply choosing different reaction temperatures. 3:1 mixture of sulphuric acid:nitric acid refluxing for oxidation of graphite/CNT also produces nearly identical CDs of 3–4 nm. For separation of CDs the standard method of dispersal of mixture in water and then filtration followed by dialysis is done to remove CNT [54]. SWCNT is also used as precursor. It is first etched by refluxing with nitric acid, which

yields CDs with some disorders, then it is *hydrothermally oxidized* by suspending it in water and heating it at high temperature (200 °C) [55].

- Acidic oxidation of micro-sized pitch-based carbon fibers that have stacked graphitic zones also produces CDs. Peng *et al.* [53] have shown that these CDs are of 1–4 nm size, 1–2 atomic layers thick and most of them have zig-zag edges.

1.4.4 Solvothermal Method

Solvothermal method is not much different from hydrothermal method. Though it is a cost-effective, eco-friendly and non-toxic method, it has poor control over sizes. In this method, solution of organic precursor is sealed and reacted in a hydrothermal reactor at high temperature to obtain CDs. For example, precursor such as glucose in water–glycol medium is heated at 180 °C in the presence of nitric acid to get CDs, which is later modified with polyethyleneimine (PEI) to improve its fluorescence quality. Precursors like glucose, citric acid, chitosan, banana juice and protein have also been used to synthesize CDs by solvothermal method.

1.4.5 Laser Ablation of Graphite

This technique is generally used to synthesize inorganic type of nanoparticles in which a solid substrate is used, on which high energy is delivered by a laser beam, which causes removal or ablation of particulates in nano form. The precursors tried for laser ablation were mostly graphite, along with different medium such as:

- ***Graphite mixed with cement using water***: Just after the discovery of CDs, Sun *et al.* [56] demonstrated synthesis of CDs using a similar method by ablation of carbon target (mixture of graphite and cement) in the presence of water vapor and argon as carrier gas at 900 °C and 75 kPa. CD aggregates of various sizes were obtained but did not show any photoluminescence even after acidic treatment. However, surface passivation with simple organic molecules successfully induced fluorescence in CDs. This was one of the very first reports to actually call carbon nanoparticles CDs, which also opened a new direction of surface passivation for enhancing fluorescence.

- ***Graphite dispersed in PEG solution:*** Hu and colleagues [57] dispersed graphite flakes in PEG_{1500N} solution for laser ablation, which yielded a homogeneous dark-colored solution. The interesting point of this method was that ultrasound was used in combination with ablation. The CDs were purified by centrifugation. The quantum yield is influenced by size variation, which in turn can be controlled by regulating the pulse width of the laser.
- ***Graphite immersed in deionized water:*** Gonçalves *et al.* [58] demonstrated the synthesis of carbon nanoparticles by laser ablation of carbon targets immersed in deionized water. The as-obtained nanoparticles were not found to be fluorescent in nature, similar to other research using the same method. The nanoparticles were refluxed in nitric acid solution for 12 hours for surface activation followed by treatment with PEG_{200} and mercapto-succinic acid. Finally, the CDs exhibited fluorescence property.

The laser ablation method possesses advantages like functionalization of CDs to make them more compatible for biological applications, as well as helps with the detection purpose of various molecules. But at the same time, it also has disadvantages, such as complex experimentation which requires multiple steps, as well as surface functionalization being a necessity in most of the cases for further applications. The method also utilizes an immense amount of energy for producing laser.

1.4.6 Pulsed Laser Irradiation of Carbon Source

Pulsed laser irradiation is a good method for synthesizing graphene quantum dots (GQD), which are also carbon QD. The basic difference between GQD and CDs is that the carbon dots are spherical single-layer structures, whereas GQDs are planar nano flakes with a few atomic layers, thus having higher surface-to-volume ratio than spherical CDs. Benzene has been successfully used in this method using nickel oxide as catalyst and irradiating them with 1064 nm pulsed Nd:Yag laser resulted in formation of GQD. Nickel was separated by centrifugation and the remaining benzene is separated by rotor evaporation. This process gives good yield of water-soluble 2.6 nm GQD across three layers [59].

Nanocarbon dispersed in solvent, such as ethanol, acetone, or water, sonicated and irradiated with Nd:YAG pulsed with a second harmonic wavelength of 532 nm also yields good quality CDs [29]. In this process the suspension of nanocarbon has to be taken into a glass cell for laser irradiation. The experimental setup is described in Figure 1.5.

1.4.7 Arc Discharge Method

As reported by Xu *et al.* [25], CDs were accidently discovered during purification of SWCNTs obtained from arc discharge soot. Synthesis of SWCNT by arc discharge method is a well-known process. During this process some fragments of carbon material (soot) are also formed. To separate and purify the CDs from soot, it was oxidized with 3.3 M HNO_3, which induced carboxyl group formation. The sediment was then extracted with NaOH (pH 8.4) solution, which resulted in the formation of a stable black suspension which was further purified/separated by gel electrophoresis. The FTIR spectrum of the fast-moving band of highly fluorescent CDs had an average size of

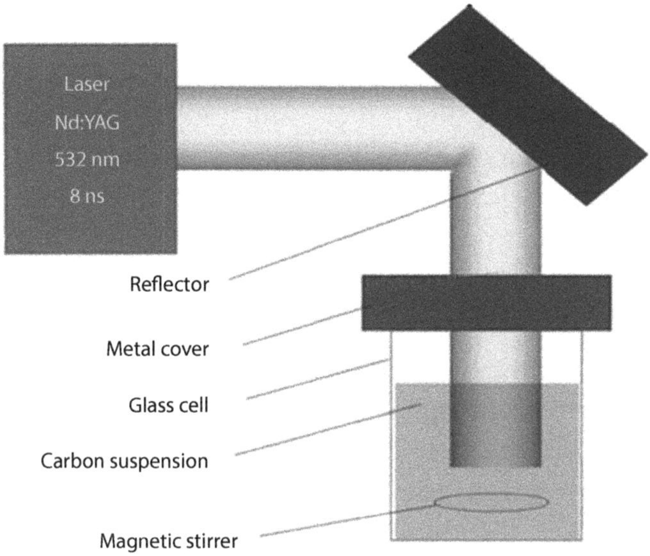

Figure 1.5 Pulsed laser irradiation of nanocarbon source dispersed in solvent to synthesize CD. (Reproduced with permission from [29])

18 nm and displayed oxygen-containing functional groups on the surface.

In a similar report, fluorescent nanoparticles were isolated from pristine and nitric acid oxidized CNTs, which were produced by electric arc discharge method [60]. The isolated fluorescent nanoparticles were hydrophobic in nature.

The arc discharge method is not widely used because the nanotubes extracted from arc soot are highly complex and occur with unwanted carbonaceous materials. Also, the CDs produced via this method require a mandatory step of purification [61, 25].

1.4.8 Plasma Treatment

A one-step synthesis and simultaneous functionalization of CDs has been achieved using submerged Plasma Arc Reactor [62]. Helium atmosphere was used and simultaneous functionalization of CDs was carried out with ethylenediamine.

1.4.9 Opening of Fullerene Cage

Loh *et al.* [63] developed this mechanistic approach for the synthesis of CDs. In this method C_{60} fullerene is opened by metal catalysis at high temperature. The fragmentation of the embedded molecules at elevated temperatures produces carbon clusters that undergoes diffusion and aggregation to form CDs.

1.4.10 Ultrasonication Method

In this process ultrasonic waves give rise to alternation expansion and compression in the reaction mixture. This in turn gives rise to interparticle collisions, which changes the morphology, composition, reactivity and properties of the articles. Ultrasonic waves in combination with carbonization of the precursors lead to the formation of CDs. Standard carbon sources that have been used are:

- ***Glucose + Acid or Alkali:*** In one of the reports, synthesis of CD from ultrasonication of glucose along with acid or alkali [64] yields water-soluble and spherical CDs, which displays NIR emission. This is one of the very important properties, which can be utilized in photothermal therapy of cancer.
- ***Carbon Source + Hydrogen Peroxide:*** CDs can be synthesized by ultrasonication of carbon source in the presence of hydrogen peroxide [65]. Thus-synthesized CDs displayed strong photoluminescence properties. FTIR analysis revealed the surface to be coated with hydroxyl and carboxyl groups, which made the CDs water soluble.
- ***Ultrasonication in combination with electrochemical method:*** Synthesis of CDs at room temperature incubation in combination with ultrasonic method using sugarcane juice as a natural carbon-containing precursor has been reported by Sharon's group [66]. This method did not need

external high temperature to initiate carbonization. In a typical procedure the purified sugarcane juice was subjected to sonication in the presence of NaOH and ethanol, which were used as surface passivation agents. The solution was stored at room temperature covered for 45–60 minutes. The dark brown colored solution was centrifuged and dialyzed against nanopure water to get CDs. The CD solution showed bright green fluorescence under UV light and was found to be stable for long periods without showing any photobleaching. The only disadvantage to this ultrasonic method is that the CDs synthesized can vary in shape and size. Also, the mixture contains CDs along with other unwanted carbonaceous material, which makes the purification step very important.

1.4.11 Microwave-Assisted Method

Microwave-assisted method has become the most widely used technique, after hydrothermal synthesis, because of its rapid heating and the resulting dramatic increase in reaction rate, cost effectiveness and simplicity. It is more convenient than other techniques because of its ability of quickly heating the carbon-containing precursors, which in turn leads to high yields of CDs in a short time. CDs produced by a microwave method showed bright luminescence when their surfaces were passivated with diamine-terminated oligomeric poly(ethylene glycol) [67]. An additional advantage is that CDs are produced in a significantly shorter time. The excellent heating capacity of microwave irradiation lies in the fact that it can heat substances uniformly. It is believed that microwave irradiation produces a force on charged particles and causes them to move or become further polarized. This leads to friction and collisions of molecules, which subsequently generates heat. Rapid heating by microwave irradiation eliminates the effect of a temperature gradient, contributing to the monodispersion, small size and high purity of the resulting CDs. The microwave-assisted pyrolysis approach to synthesize fluorescent CDs with electrochemiluminescence (ECL) properties is a rather cheap process.

- Using ***citric acid and urea*** as carbon precursors, Qu *et al.* [68] have demonstrated the synthesis of CDs via one-step microwave-assisted heating. In a typical experiment, the colorless aqueous solution of citric acid and urea is heated in a microwave for approximately 4 minutes followed by heating in a vacuum oven at 60 °C to obtain a dark brown colored residue. It was purified by centrifugation to yield fluorescent CDs in the size range of 1–5 nm.
- Similarly, using a mixture of ***aqueous solution of citric acid with 2-ethylenediamine***, CDs can be prepared by microwave-assisted heating [69]. The obtained CDs exhibit excitation-dependent fluorescent spectra. FTIR analyses revealed the presence of carboxyl and amine groups, which lead to fluorescent properties.
- A mixture of ***calcium nitrate and urea*** is used to produce nitrogen-doped CDs [70] by microwave-assisted heating.
- Sharon's group have used aqueous solutions of ***sorbitol*** as well as ***phenylalanine*** to obtain CD by subjecting these precursors to microwave heating with addition of 1M NaOH and ethanol mixture; both precursors yielded crude brown solution. The crude solution mostly contained CDs as well as other carbonaceous materials and hence was subjected to dialysis. Pure solution of CDs was obtained which showed blue fluorescence under UV light (365 nm) for CDs obtained from phenylalanine precursor [71]. ***NaOH and ethanol*** were

used during the process in order to surface passivate the CDs with oxygen-containing functional groups. The purified CDs were used further for drug delivery applications.

- Sharon's group also employed natural plant-derived precursor, i.e., ***Acacia arabica* gum**, for synthesis of CDs by same microwave-assisted method [72].
- ***Ethylenediamine and Phytic acid mixture*** were processed by a one-step microwave-assisted heating method [73]. The turbid solution obtained after microwave heating was purified to get phosphorous-containing CDs. The CDs displayed carboxyl and hydroxyl groups on their surface, which increased their stability in aqueous solution. Also, no extra passivation step was required for inducing fluorescence. The highest quantum yield of CD was calculated to be 21.65%.
- Wang *et al.* [74] suggested that in this microwave-assisted heating process, a solution of ***polysaccharides, glycerol, glycol, glucose, sucrose, etc.***, can be used as carbon source without any passivating agent and will require only a few minutes of reaction time to form CDs. It has been noted that with increasing reaction time, color of the solution changes from colorless to yellow and finally to dark brown, which implies the formation of CDs.
- A microwave-assisted method in combination with hydrothermal method using glucose can also produce water-soluble crystalline CDs with an average diameter as small as 1.65 nm. This method does not require any surface passivation agents or inorganic additives. CDs can be produced just by pyrolyzing glucose, sucrose or fructose [29].
- CDs can also be fabricated by ultrasonication of glucose or active carbon. Such monodispersed water-soluble fluorescent CDs were found to emit

bright and colorful PL covering the entire visible-to-NIR spectral range.
- Some research is also being done to create QD by changing the width of GNRs at select points along the ribbon, creating quantum confinement [27].

Even though this method is widespread for synthesis of CDs because of its various advantages, it also carries a few disadvantages like the quantum yield of CD is slightly lower as compared to other techniques and also that surface passivation and purification steps are required for enhancing their properties.

1.4.12 Chemical Methods

Various chemical approaches have been employed for synthesis of CDs. Some of them are mentioned here. Solution phase chemical methods by ***oxidative condensation of aryl groups*** have been successfully applied to produce CDs. Versatile synthesis of large, stable colloidal CDs with desired sizes and structures ***via a new solubilization strategy*** was demonstrated by Li *et al.* [29]. ***Un-substituted hexa-perihexabenzocoronene (HBC)*** is used as a carbon source to synthesize multicolor CDs of uniform size of 6 nm diameter and 2–3 nm thickness. This method has produced the largest CDs reported so far.

A method employing reverse micelles as nanoreactors is also used to produce highly luminescent CDs [75]. CDs were produced ***via glucose carbonization in reverse micelles*** followed by *in-situ* surface passivation, as shown in Figure 1.6. This solution-phase chemical method offers size tunability.

1.4.13 Supported Synthetic Procedure

The supported synthetic method, which is also sometimes called template-based method, is generally used to obtain monodispersed nanoparticles involving molecular sieve, porous carbon and so on. The host material serves as a support and provides an environment to synthesize CD and also prevents their agglomeration.

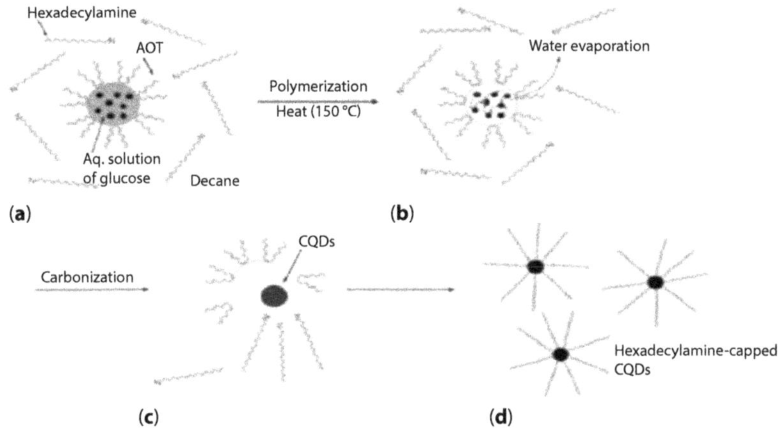

Figure 1.6 Schematic representation of the synthesis of CDs in reverse micelles: (a) formation of water-in-oil reverse micelles, (b) hydrolytic polymerization, (c) carbonization and (d) *in-situ* passivation by hexadecylamine. (Reproduced with permission from [75])

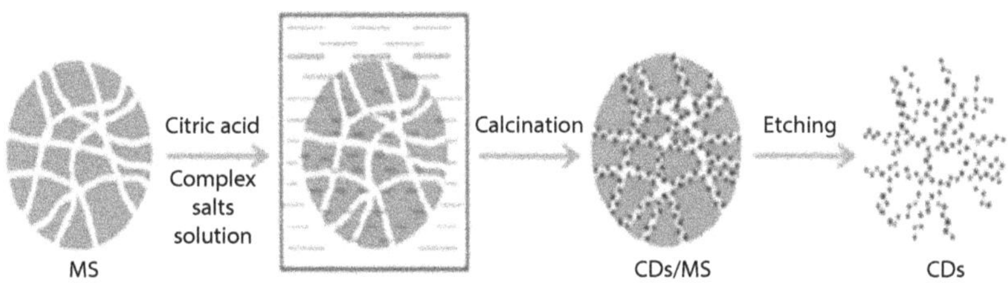

Figure 1.7 Schematic representation of synthesis of CDs using mesoporous silica as host particles. (Reproduced with permission from [76])

It is a simple procedure to synthesize CDs using mesoporous silica spheres as host or nanoreactors developed by Zong *et al.* [76]. In this process, first mesoporous silica spheres are synthesized using tetrahexadecylamine as a surfactant and tetraethoxysilane as precursor in the presence of ammonia as a catalyst (Figure 1.7). The spheres are then used as nanoreactors for synthesis of CDs using citric acid as carbon precursor. The mentioned mixture along with complex salt solution is subjected to repeated calcinations to obtain CD and mesoporous sphere composites. The mesoporous spheres are etched using NaOH to get free CDs. The use of spheres with nanopores not only helps in confining the size of CDs but also prevents their aggregation. It is interesting to note that the CDs were found to be highly photoluminescent (PL) without any further treatment.

The CDs can also be synthesized with the support of silica spheres and resols as carbon precursors [77]. The procedure added surfactant modified spheres to carbon precursor. During the heating step the resols polymerized on the surface of spheres, which on further thermal treatment at high temperature in the presence of argon gas produces CD-sphere composite. Lastly, the composites are etched with NaOH solution to obtain CDs in the size range of 1.5–2.5 nm. These CDs are uniform in size and non-aggregated and further surface passivated using HNO_3 and PEG_{1500N} separately on 2 batches to enhance PL emission. Interestingly, in this whole procedure the surfactant serves as a template for polymerization of carbon precursors, as well as prevents the agglomeration of CDs.

With a slight variation, CD is synthesized using NaY zeolite as the support/host [78], using ion exchange process in co-ordination with thermal oxidation followed by etching with strong acid to get fluorescent CDs. The CDs obtained by this method are spherical with size ranging from 4–6 nm.

Another method of synthesis of CDs from glucose involves the use of reverse micelles as nanoreactors [75]. During the process glucose undergoes carbonization and hydrolytic polymerization. Surface passivation also takes place simultaneously and highly fluorescent CDs are produced.

The only disadvantage to using the supported route is that the high temperature treatment of host particle leads (in the case of mesoporous silica particles) to an increase in structural density, which makes the removal of host matrix by etching difficult.

1.4.14 Biogenic Method

Biogenic synthesis of CDs involves the use of plant parts or plant-derived materials as precursors. Some of the successfully used precursors are orange juice, jaggery, bread, sugar, sugarcane juice, milk, *Trapa bispinosa* peel extract, etc., to synthesize highly fluorescent CDs. In short, it can be said that by carbonizing the rich carbon sources and in some cases by passivating the surfaces, photoluminescent and chemically inert CDs are synthesized.

- Using size-dependent separation by sucrose density gradient centrifugation, Sharon's group [66] successfully synthesized luminescent water-soluble CDs using water plant ***Trapa bispinosa peel extract*** without adding any external oxidizing agent at 90 °C. These CDs showed prominent green fluorescence under UV light (λ max). UV-Vis spectra of these CDs were recorded at different time intervals (30–120 min). The data obtained displayed signature absorption of CDs between 400 and 600 nm. Fluorescence spectra of the dispersion were recorded after 120 min of synthesis. These spectra exhibited characteristic emission peaks of CDs when excited at 350, 400, 450 and 500 nm. X-ray diffraction (XRD), Raman spectroscopy and thermo-gravimetric analysis (TGA) were used to further analyze the CDs. The XRD revealed the structure of the CDs to be turbostatic. CDs produced by this method were found to be exceptionally biocompatible with MDCK cells.
- Sharon's group [72] was able to synthesize hydrophobic CDs from ***Gum arabic*** by microwave-assisted methods in the presence of ethanol and NaOH within 4 min. The CDs were extracted using chloroform. The PL spectrum at different excitation wavelengths of these CDs showed red-shifted pattern. The excitation wavelength-dependent emission wavelength and intensity were attributed to quantum effects and/or emissive traps

present on their surface owing to the presence of functional groups. Partial self-passivation leads to variation in sizes; however, they remained spherical. The CDs had –OH groups, C–H groups, carbonyl, C–O and C–O– groups. The plane of the carbon has poor crystallinity or nearly amorphous nature. Although the synthesized CDs appeared as a brownish-black precipitate in water, they were immediately extracted into organic solvents and therefore the hydrophobicity of such CDs was illustrated simply with the ease of dispersion in different organic solvents. The mechanism of formation of hydrophobic CD was probably by fragmentation of gum arabic forming the carbon core, followed by mild partial passivation via condensation with hydroxyl groups of fragmented or un-fragmented gum under microwave heating conditions forming C–O–C bonds, which is known to occur with active hydroxyl end groups. Remaining marginal hydroxyl and carbonyl groups on the surface of those hydrophobic CDs resulted in the formation of defect sites and emissive traps, which contributed to its PL properties.

- Himaja *et al.* [79] have demonstrated the synthesis of CDs using kitchen wastes. A fine paste was prepared from peels of fresh cucumber and pineapple using water. The precursors were purposely used in a paste form so that maximum carbon can be extracted. The peel paste was mixed with distilled water and subjected to heating at 150 degrees for 2 h. The solution was then filtered using a simple house filter and was centrifuged at 4500 rpm for 20 min to remove larger particles. The pale-yellow solution obtained was subjected to refluxing at 150 degrees for 20 min followed by suspension in 1 N NaOH.

The final solution so obtained displayed green fluorescence under UV light. This work shows a very simple and cost-effective approach for large-scale green synthesis of biocompatible CDs.

- Sahu *et al.* [80] have synthesized CDs using orange juice via a simple one-step hydrothermal procedure. The CDs were mixed with ethanol and heated at a constant temperature of 120 degrees in an autoclave for 120 min. The cooled dark-colored mixture was then treated with dichloromethane to wash out organic moieties. The aqueous solution was subjected to centrifugation at 3000 rpm for 15 min. The pellet obtained was identified as larger and had less fluorescent particles, whereas the brown supernatant was centrifuged at 10,000 rpm for 15 min to yield highly fluorescent and small CDs (in size range of 1.5–4.5 nm). These CDs gave a quantum yield of 26%, which is excellent in comparison to semiconductor QD.

It is very interesting to know that the majority of the methods which used a natural precursor for synthesis of CDs did not require any further surface passivation. Also, these CDs displayed higher biocompatibility as compared to chemical precursor-based CDs. The only disadvantage of using a natural precursor was that the chemical composition of the CDs cannot be controlled.

1.5 Characterization of Carbon Dots

There are various techniques of CD characterizations. This section explains all possible characterization techniques which can be or have been used for identifying and measuring CD properties.

1.5.1 Microscopic Methods

The following microscopic techniques are used for physical morphological studies of CDs.

1.5.1.1 SEM and TEM Characterization

The morphological survey of CDs can be performed using electron microscopy techniques; SEM (scanning electron microscopy) and TEM (transmission electron microscopy). In SEM, the image is formed depending on the number of secondary electrons (reflected from sample under study) detected by electron detector. There are various methods to capture a perfect image using a SEM apparatus. The CD sample for SEM analysis is mostly prepared by spreading C-dot solution on a substrate (e.g., silicon wafer) followed by metal coating of a few (2–3) nanometers for better secondary electron output [81], resulting in more clear imaging of CDs as compared with CDs imaged without coating of metal. Figure 1.8 shows a schematic diagram of secondary electron generated with and without metal coating of CDs.

On the other hand, TEM analysis is a more useful technique because it provides more detailed morphological analysis of CDs with the resolution of a few nanometers. For sample preparation, a few drops of diluted CD solution are deposited onto amorphous carbon-coated copper grid (or any other grid depending upon availability and convenience), followed

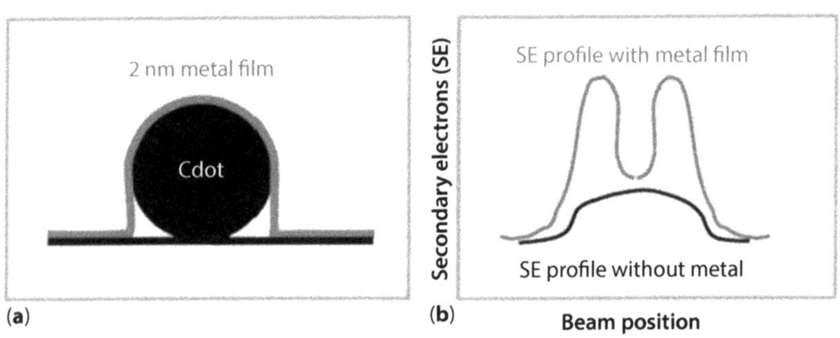

Figure 1.8 Schematic representation of (a) CD coated with metal film and (b) corresponding secondary electron profile.

by evaporation to dry. The TEM image of sample is formed relative to amorphous carbon contrast on background. TEM analysis provides clear imaging of CDs along with information related to interatomic distance of carbon atoms (Figure 1.9). This information helps in finding the crystalline nature of CDs.

Figure 1.10a shows spherical- as well as oval-shaped CDs in the size range of 5–10 nm. The crystalline nature is also confirmed with electron diffraction pattern. Selected area electron diffraction (SAED) pattern measured by TEM, as shown in Figure 1.10b, demonstrates the amorphous nature of CDs synthesized using neem gum at room temperature employing ethanol and sodium hydroxide as surface passivating agents [83]. Energy-dispersive spectroscopy (EDS) and electron energy loss spectroscopy (EELS) can also be done using TEM. These characterizations are useful in finding out the elemental composition of CDs. With the help of SEM and TEM, various information about CDs, e.g., surface morphology, crystalline nature as well as elemental composition, can be determined and this information can help in practical applications of CDs in a variety of fields.

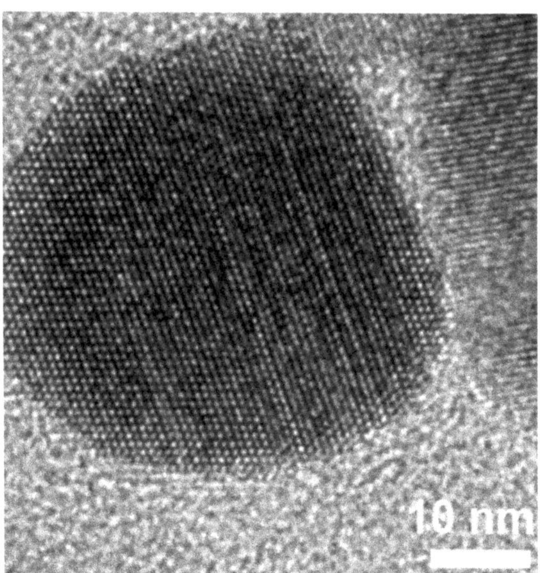

Figure 1.9 TEM image of Carbon Dot.

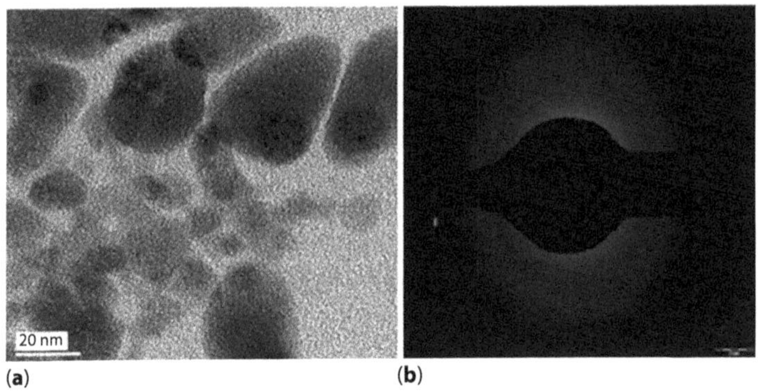

Figure 1.10 (a) TEM image of CDs obtained from neem gum and (b) SAED image of neem gum CDs (Reproduced with permission from [83])

1.5.1.2 AFM and STM Characterization

Atomic force microscopy (AFM) and scanning tunneling microscopy (STM) are types of scanning probe microscopy (SPM) used to determine the size of CDs and surface morphologies to the resolution of a fraction of nanometers. In AFM, a silicon cantilever containing a sharp tip is used to scan all over the surface of substrate upon which the CDs are dispersed. Atoms of the cantilever tip and sample surface exert force on each other; this force is measured with respect to the different heights of the sample surface during the scan, and this scanning in turn generates a height/position profile of the CDs with reference to the substrate. Another measurement type is detection of laser light deflection from the backside of cantilever during the scan. Even a slight change in cantilever tip height during the scan deflects laser and this deflection is recorded by photodiode, amplified and coded to create an image of the sample under study. This measurement type is high resolution and easier as compared to the former. Figure 1.11 shows the AFM image of CDs. The image shows that CDs are evenly distributed with an average size of 5 nm. The line profile (below the AFM image) measured at the white line highlighted in the AFM image also confirms the same [84].

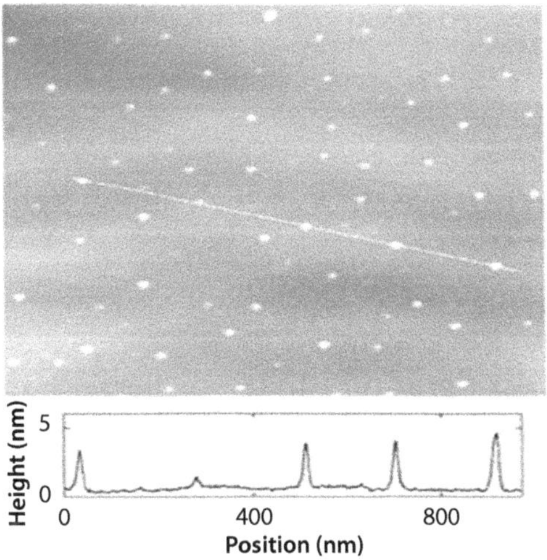

Figure 1.11 AFM image of CDs.

Another technique similar to AFM is STM, but due to a continuous voltage between the STM probes called "tunneling voltage," this voltage causes tunneling of electrons between atoms of sample surface once brought very close to each other. These transmitted electrons in turn cause a change in current probe and this change in current with respect to height is used to form the STM image.

1.5.2 Spectroscopic Methods

Atoms and molecules interact with electromagnetic radiation (EMR) in a wide variety of ways. Atoms and molecules may absorb and/or emit EMR. Absorption of EMR stimulates different types of motion in atoms and/or molecules. The patterns of absorption (wavelengths absorbed and to what extent) and/or emission (wavelengths emitted and their respective intensities) are called a "*spectra*." The field of spectroscopy is concerned with the interpretation of spectra in terms of atomic and molecular structure (and environment).

A standard spectrophotometer instrument consists of the following parts (Figure 1.12):

- Radiation Source S.
- Vertical entrance slit S_1, which allows light into the instrument.
- Spherical Mirror M_1, collimates a parallel beam of light onto the grating.
- Diffraction grating that disperses the individual components of the incident white light and delivers a parallel beam onto the spherical mirror M_2.
- As the grating is rotated, a spherical mirror M_2 receives the wavelength components from the grating and reflects as a focused beam at exit slit S_2.
- The light signal exiting from the instrument is made incident on a spherical mirror M_3.
- M_3 focuses the incident signal on a detector.
- Finally, the signal is amplified and processed to display the results.

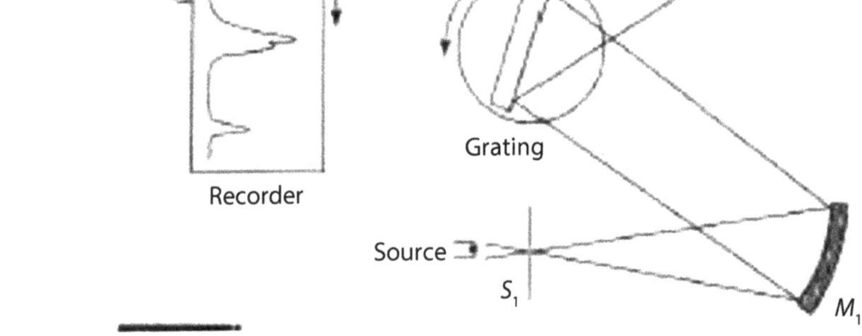

Figure 1.12 Schematic diagram of a spectrophotometer.

1.5.2.1 UV-Vis Spectroscopy and its Application for Band Gap Determination

UV-Vis spectroscopy is used to determine the excitation wavelengths and the absorbance spectra of CDs. The π-electrons or non-bonding electrons (n-electrons) of molecules under study absorb the energy of ultraviolet or visible light depending upon energy gap up to higher anti-bonding molecular orbitals to excite. There are four possible transitions, namely, π-π^*, n-π^*, σ-σ^*, and n-σ^*. The energy required for various transitions are in the order σ-σ^* > n-σ^* > π-π^* > n-π^*. The CDs generally absorb in the UV region with a feeble extending low intensity tail in the visible region. CDs also typically show excitation-dependent emission spectra, which gives rise to fluorescence which can be detected under UV light.

A representative UV-Vis spectra of CD is displayed in Figure 1.13 [85]. The UV-Vis spectra (Figure 1.13a) shows a wide range of absorption ranging from approximately 200 nm up to 450 nm and further extending flat tail in the visible region. The highest absorption peak is observed at 275 nm, which arises due to the π-π^* transitions of sp^2 carbon. The emission spectra show a continuous shift from 398 nm to 454 nm when excited in the range of 300–400 nm (Figure 1.13b). The maximum emission intensity was observed at 440 nm when the sample was excited at 365 nm. Thus, excitation-dependent emission spectra are observed, which is one of the very important optical properties.

1.5.2.2 Fluorescence Spectrometry

Fluorescence spectrometry is similar to UV-Vis spectroscopy. It can be used to characterize CDs or any other nanoparticles for determination of size and concentration based on its fluorescence properties. In principle, a beam with a variable wavelength (200–800 nm) is used to illuminate the sample solution with CDs in a cuvette from which the emitted light (of another wavelength) is measured. In fluorescence spectrometry both an excitation

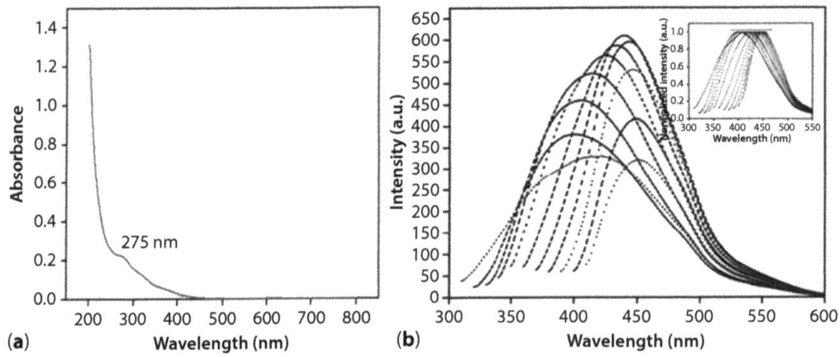

Figure 1.13 (a) UV-Vis absorption spectra of CD and (b) Emission spectra when excited from 300–400 nm. Inset is normalized emission spectra. (Reproduced with permission from [85])

spectrum and/or an emission spectrum can be measured. The concentration of sample (CDs in this case) in the solution is directly proportional to the intensity of the emission spectra.

Mao *et al.* [86] synthesized CDs using lamp black and studied the fluorescence properties of as-prepared as well as surface-passivated CDs. Fluorescence intensity and lifetime measurements were carried out on a Fluorolog-3 fluorescence spectrometer (Horiba Jobin Yvon Inc., France). Excitation wavelengths of 200 nm and 460 nm were used, whereas emissions at 520 nm were collected. The maximum excitations were observed at 470 nm and maximum emissions were noted at 520 nm. The fluorescence lifetime of passivated CDs was found to be longer as compared to bare ones (Figure 1.14).

1.5.2.3 Fourier Transform Infrared (FTIR) Spectroscopy

Spectral evidence to find the functional groups over the surface of CDs is provided by FTIR spectroscopy. Usually, hydroxyl, carboxyl, ether or epoxy and other oxygen-containing functional groups are attached to the surface of CDs during the synthesis process or due to intentional surface passivation treatments. CDs have applications in drug delivery, bioimaging, as well as sensors in which various drugs, labels or ligands have to be intentionally attached to the surface of CDs or, in

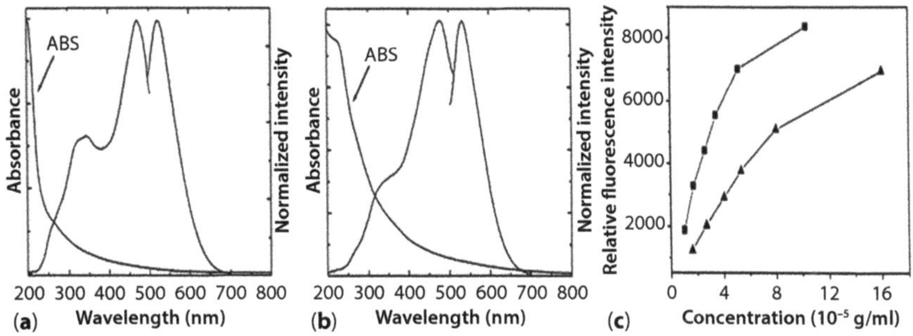

Figure 1.14 Absorption and fluorescence spectra of (a) Bare CDs, (b) Passivated CDs and (c) Plot of fluorescence intensity of bare CDs (▲) and modified CDs (■) vs. concentration. (Reproduced with permission from [86])

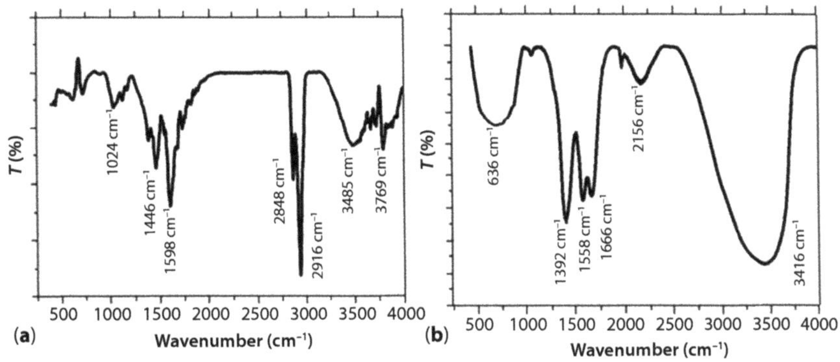

Figure 1.15 FTIR spectra of (a) Biological CDs and (b) Ciprofloxacin attached CDs. (Reproduced with permission from [72])

other words, CDs need to be modified with respect to particular applications. Hence, in order to confirm these attachments (passivation) FTIR acts as a very important characterizing technique. FTIR, which works on the principle of Michelson interferometer, is also a key characterization method in determining the cause of the optical properties of CDs. For example, in Figure 1.15 FTIR analyses of CDs synthesized using the natural precursor gum arabic is shown [72]. The CD sample was analyzed within the spectral window of 500–4000 cm^{-1}. The comparative FTIR spectra of CDs and ciprofloxacin-loaded CDs shows self-passivated CDs containing C-N stretching,

-CH bending and C=C aromatic ring stretching. Other related groups found were hydroxyl and amine stretches arising due to the solvent and primary amines respectively (Figure 1.15a). Figure 1.15b displays the FTIR spectrum of CDs conjugated with ciprofloxacin. The graph shows significant changes, like addition of a new peak due to new –CH bends, as well as shift in peaks because of the possible interaction of –NO groups on the drug with the –OH groups on CDs. All these changes confirm the chemical interactions forming weak hydrogen bonds as well as strong amide linkages.

1.5.2.4 X-Ray Diffraction (XRD) Analysis

To determine the state of crystallinity of carbon dots or graphene QD, XRD is used. In XRD, an X-ray beam is bombarded on the sample surface, which in turn forms a diffraction pattern. The pattern contains the information of interatomic distance of carbon atoms, revealing the state of crystallinity of the materials under study. A sharp peak at 26° (2θ) angle in XRD pattern indicates a highly crystalline carbon, whereas a broad peak around this angle indicates amorphous carbon. Figure 1.16 displays a typical XRD graph of CDs showing a clear sharp peak

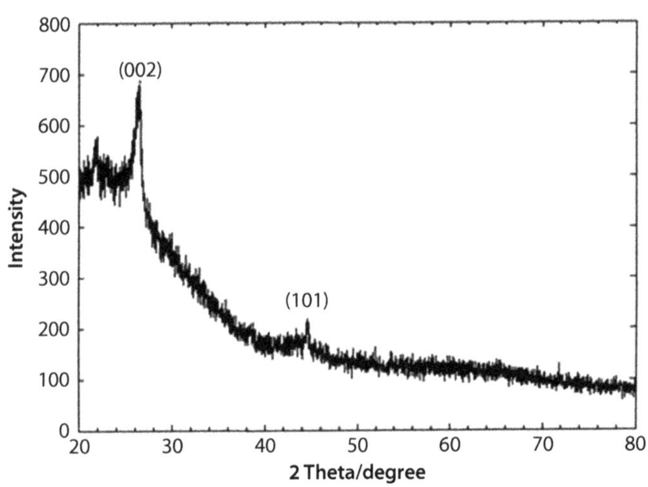

Figure 1.16 A typical XRD graph of CDs.

at 27° (2θ) which is assigned to the (002) plane of graphitic carbon, whereas 44° (2θ) is assigned to the (101) plane of graphitic carbon.

1.5.2.5 X-Ray Photoelectron Spectroscopy (XPS)

The XPS characterization technique is used to find out the types of bonding states on the surface of CDs. As mentioned earlier, the surface of CDs is rich in various functional groups. These groups and CDs' composition can both be characterized using XPS. Usually XPS and FTIR complement each other. XPS uses high-energy X-rays to bombard the sample surface under study and this high-energy radiation causes emission of various types of electrons, those electrons which are emitted from the innermost shell of atoms usually carry high energy and these electrons are detected in electron detector and plotted on an energy plot against number of counts (Figure 1.17). XPS usually can only measure sample surface up to a depth maximum of 5 nanometers.

For example, hydrothermal synthesis of CDs using lignin in the presence of hydrogen peroxide is demonstrated by Chen *et al.* [44]. The XPS survey of CD shows prominent carbon and oxygen peaks at 285.5 and 532 eV respectively (Figure 1.17a). The deconvolution of C1s and O1s are shown in Figure 1.17(b) and 1.17(c) respectively. The C1s plot shows the presence of sp^2 carbon (283.2eV), sp^3 carbon (286.1 eV) and carboxyl carbon (288.6 eV) and the Os1 spectra exhibited three peaks which are assigned to C=O (530.6 eV), C-OH (532.1 eV) and C-O-C (533.2 eV) groups.

1.5.2.6 Dynamic Light Scattering/Photon Correlation Spectroscopy (DLS/PCS)

Dynamic light scattering (DLS), sometimes referred to as photon correlation spectroscopy, is a well-established technique for measuring the size and size distribution of molecules and particles typically in the range of a few nanometers. Typical

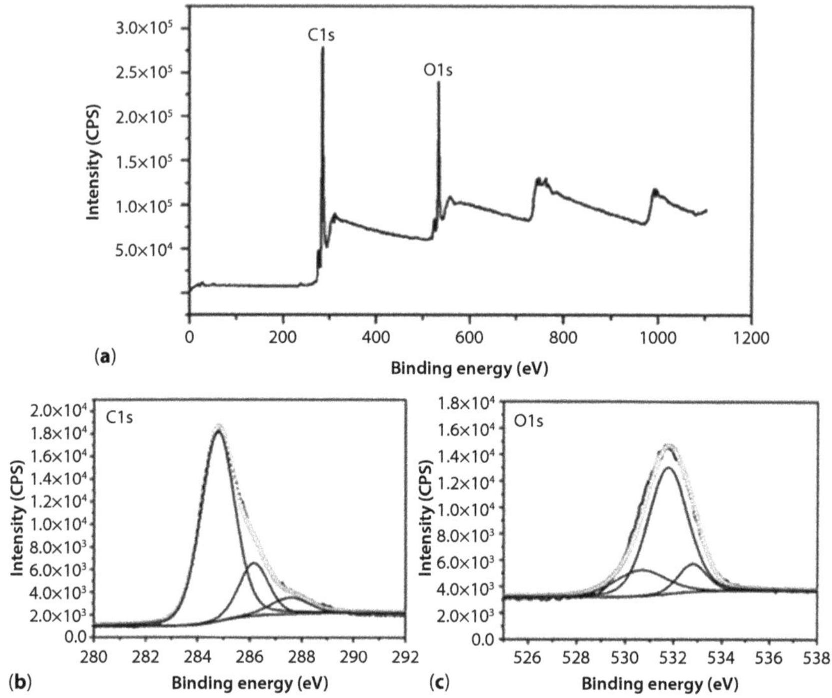

Figure 1.17 (a) XPS, (b) C1s and (c) O1s spectra of as-prepared CDs. (Reproduced with permission from [44])

applications of dynamic light scattering are the characterization of particles/molecules, which have been suspended or dissolved in a liquid. The Brownian motion of particles (molecules) in solution causes laser light to be scattered at different intensities and the careful investigation of these intensity fluctuations yields the velocity of the Brownian motion, and hence the particle size using the Stokes-Einstein relationship can be easily determined.

1.5.2.7 Dual Polarization Interferometry (DPI)

Dual polarization interferometry (DPI) is an analytical technique providing real-time measurements of molecular layer dimensions along with determination of both structural and functional aspects (Figure 1.18). Using DPI, molecular

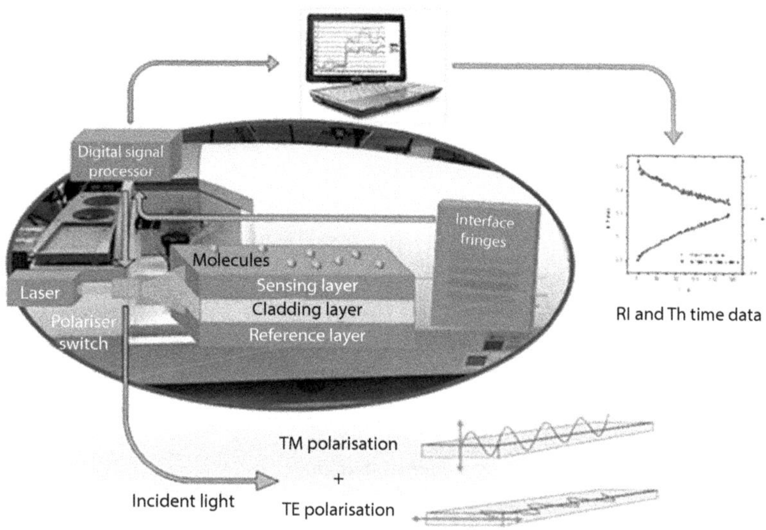

Figure 1.18 Schematic representation of DPI. (Reproduced with permission from [87])

immobilization, orientation, assembly, interactions, and stability can be examined, and can be used in determining biomolecular interactions, drug discovery, sensor surface design, biocompatibility, etc.

Platt *et al.* [87] used DPI to measure the interaction of a copolymer immobilized on the chip with antibodies acting as a mimic of the allergen-reactive immunoglobulin E and also measured the interaction of QD with streotavidin. DPI has been also used for quantum dot layer thickness measurement.

1.5.2.8 *Raman Spectroscopy*

Raman spectroscopy is a nondestructive method of finding out the state of carbon in CDs or graphene QD. The characteristic disordered sp^2 carbon peak is usually centered at 1350 cm^{-1}, called the D peak, and crystalline graphite carbon peak is centered at 1580 cm^{-1}, called the G peak. The ratio of these two peaks decides the extent of crystallinity of carbon dots. The D peak, however, may also arise to the edge effect of graphene QD or the hetero atoms attached on the surface of carbon dots.

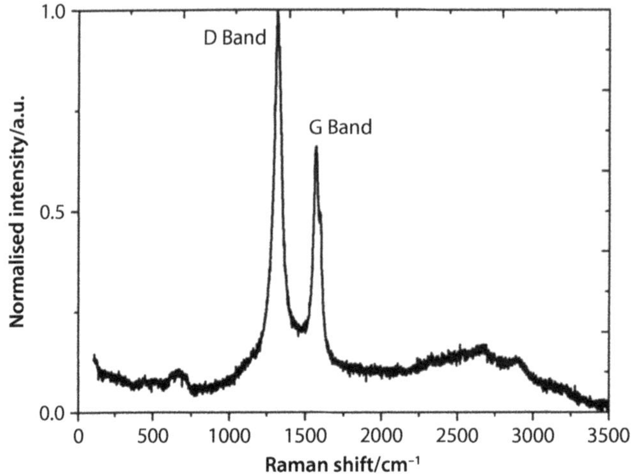

Figure 1.19 Raman spectra of CDs displaying its defective nature based on D-band and G-band intensities. (Reproduced with permission from [50])

Figure 1.19 shows the Raman spectra of purified CDs synthesized using sorbitol as carbon precursor [50]. The graph displays an intense D peak as compared to low-intensity G peak. The I_d/I_g ratio signifies the disorder in the material. The ratio of 1.42 in this case indicates the presence of pure CDs.

The functional groups on the surface of CDs can be clearly analyzed by FTIR and XPS analyses (as explained in the following sections).

1.5.2.9 Nuclear Magnetic Resonance (NMR) Spectroscopy

Nuclear magnetic resonance spectroscopy is an important technique for biological molecules. Tian *et al.* [41] have performed NMR of CDs synthesized using natural gas combustion soot in nitric acid reflux. The NMR measurements revealed that CDs were made up of sp^2 carbons. There were no signals that appeared below 120 ppm in the 13C NMR spectrum, indicating the absence of aliphatic carbons, as sp^3 carbons resonate in the range of 8–80 ppm, whereas sp^2 carbons resonate in the range of 90–180 ppm. The number of peaks which emerged within the range of 120 to 150 ppm, most likely arose from sp^2 carbons.

1.6 Summary

This chapter provides a basic understanding of quantum mechanics and concepts related to the phenomenon. The discovery, properties, applications, as well as the present drawbacks of QD is also explained. Hence, the need for a new material with enhanced properties and stability becomes the need of the hour. Ever since the discovery of CDs they have attracted interest amongst scientists and have been widely studied for their similarities as well as the surprising differences with respect to semiconductor QD. The overpowered properties in comparison to QD open up difficult as well as new fields of applications. CDs have been fabricated via different synthetic routes which are very facile and eco-friendly in contrast to that of QD. Precursors as simple as daily usable and available carbon sources like sugars, proteins, natural polymers, carbon soot, and even waste materials like bagasse, molasses and kitchen slurry, can also be used for synthesis of stable CDs. However, more and more work is still going on to acquire more stable and highly fluorescent carbon dots using the simplest techniques in order to make the procedure more cost effective. Later, various chemical strategies to enhance the fluorescence properties of carbon dots have been discussed, the most popular being surface passivation with different molecules. This procedure not only makes stable dots but at the same time also helps in increasing the quantum yield, which is one of the most important aspects for biological applications.

The second half of the chapter deals with various characterization techniques for studying carbon dots. Various structural and optical characterizations like SEM, TEM, FTIR, UV-Vis and fluorescence spectrometry, RAMAN analysis, NMR, DPI, and XRD have been discussed in detail in order to study the depth of these nanoparticles. Characterization plays a very important part because it helps in better understanding of the structure and properties of CDs and allows the design of experiments to obtain particular types of CDs for various specific applications.

2

Properties of Carbon Dots

Because the theory of quantum mechanics could explain all of chemistry and the various properties of substances, it was a tremendous success. But still there was the problem of the interaction of light and matter.

Richard P. Feynman

2.1 Introduction

A carbon dot, carbon QD or carbon nanodot (Figure 2.1) is a novel material that has emerged in the field of nanotechnology. It has garnered tremendous attention because of its many unique properties. The major unique properties of CDs are fluorescence, photoluminescence, narrow tunable emission, size-dependent photoluminescence, upconversion, large absorption coefficient and size-dependent absorption, broad excitation spectra, multiphoton excitation, good conductivity, ease of functionalization, good biocompatibility, good solubility in aqueous and polar solvents, chemical inertness, photochemical stability against photobleaching and photoblinking, biocompatibility and low toxicity. CDs exhibit nanosecond lifetimes and two-photonic cross sections similar to those of QD. These varied properties are due to the structures and components of CDs.

In this chapter the attention is paid to the properties that are useful and suitable in application of CDs as theranostic agent, i.e., for biolabeling, bioimaging and as a vehicle for drug delivery.

The CDs are very small in size and exhibit a phenomenon called *quantum confinement effect* (explained in Chapter 1), which is a reason for their optical and electronic properties. Let us understand what quantum confinement is.

We are aware of the Bohr's model of an atom and energy level associated with it. Since the electrons are defined by their principal quantum numbers (n), they have a fixed energy given to them and this leads to well-defined energy levels. The other quantum numbers define their shape, direction and spin.

When atoms come into close proximity with other atoms in a solid, most of the electrons remain localized and may be considered to remain associated with a particular atom. However, some outer electrons will become involved in bonding with

Figure 2.1 CD powder synthesized in Sharon's wcRnb Lab.

neighboring atoms. Briefly, the well-defined outer electron states of the atom overlap with those on neighboring atoms and become broadened into energy bands (valence and conduction), which are of allowed energy levels and are separated by forbidden energy regions (band gaps). These band gaps may be thought of as arising from the original energy gaps between the various atomic orbitals of the isolated atoms. When the temperature is raised the electrons in the valence band attain enough energy to overcome the band gap and enter the conduction band. When the electron is excited, it leaves behind a vacancy called a hole. This pair of electron and hole is called an exciton and the distance between them is called the Bohr's exciton radius (nm). In bulk materials this exciton is free to move in all directions.

When the size of the material is reduced to the same order as the exciton radius, i.e., to a few nanometers, constraints make the electrons "feel" the presence of the particle boundaries and respond to changes in particle size by adjusting their energy. This phenomenon is known as quantum confinement effect and the exciton properties are modified. This leads to a collapse of the continuous energy bands of a bulk material into discrete atomic-like energy levels. Quantum confinement also leads to widening of the band gap as the excitonic transition energy increases.

2.2 Optical Properties

Optical properties are the interaction of electromagnetic radiation (i.e., γ-rays, X-rays, UV rays, visible light, infrared rays and radio waves) with matter, that may result in reflection, refraction (transmitted) or absorption. During reflection some part of the reflected light could be absorbed and the other frequencies could be scattered. Absorption of light/energy activates some process in the material that takes it from the ground state to an excited state. These processes are: (i) electronic, (ii) vibrational and (iii) rotational excitations. Hence, many of the optical properties are closely related to the electrical and electronic properties of the material. Some absorbed energy could be reemitted or dissipated as heat; this is called dissipative absorption. When a wave is transmitted from one medium (vacuum) to the other, its frequency remains constant, but its velocity decreases (the wave being slower in the medium). In the next part of this chapter various optical properties that are exhibited by the CDs are discussed.

Generally, CDs show absorption in the UV region with an extending tail in the visible and sometimes also in the NIR region. Optical characters may reflect not only due to different sized particles in sample solution but also because of distribution of different emissive sites or energy traps on the surface of CDs.

For the use of CDs for diagnostic purposes, there is a need to enhance the photoluminescence that is the emission spectra (Figure 2.2). To enhance the luminescence of CDs, blue luminescent reduced state CDs were prepared by reducing CDs with $NaBH_4$. The quantum yield of CDs increased from 2–24% and maximum emission wavelength shifted from 520 to 450 nm [88].

Figure 2.2 Aqueous solution of CDs under visible light and under UV (365 nm) light showing fluorescence [synthesized in Sharon's wcRnb Lab].

2.2.1 Absorbance

Light absorption occurs when atoms or molecules take up the energy of a photon of light, thereby reducing the transmission of light as it passes through a sample. Light attenuates exponentially as it passes through clear materials or solutions. The reduction of transmitted light is exponentially related to the concentration of the sample and path length of the light traveled.

Typically, CDs show strong optical absorption in the UV region, which gradually decreases in the visible range. Generally, CDs show two peaks, one in the deep UV region and the second in the near visible region. Both peaks arise due to $\Pi \rightarrow \Pi^*$ electronic transition of C=C bonds and the $n \rightarrow \Pi^*$ electronic transitions due to carbonyl groups. But sometimes single absorption peak is also observed due to clustered structures. Figure 2.3 shows representative absorption spectra of CDs. Most of the CDs, such as those prepared by laser passivation, microwave/ultrasonic, or supported method and electrochemical oxidation have an absorption band around 260–320 nm [29]. However, the absorbance of CDs increases to longer wavelength after surface passivation [89] with different

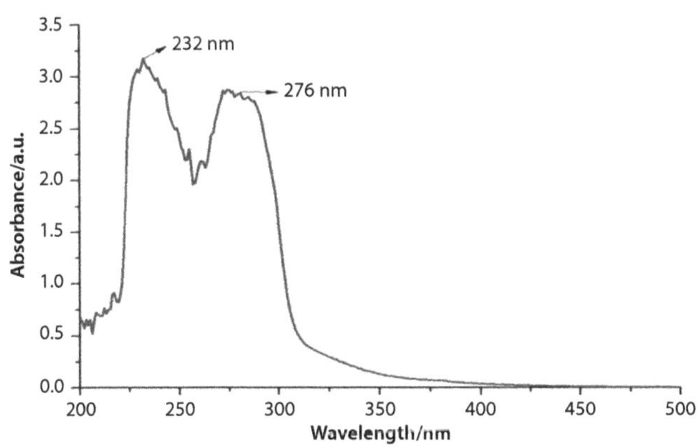

Figure 2.3 UV-Vis spectra of CDs showing characteristic absorption peak between 210–300 nm.

chemical moieties. This is due to the presence of surface emissive traps or due to quantum confinement effect. Since CD has sp^2/sp^3 hybridized carbon symmetry along with distorted carbon structures, suggesting that CDs are conjugated systems that might be the reason for their absorption in the UV-Vis spectral region.

Another example is of CDs synthesized using citric acid as carbon precursor by the process of carbonization at 200 °C [90]. This CD showed an UV absorption peak at 362 nm with a narrow width, indicating regular or uniform sized CDs. The maximum emission wavelength remained unchanged upon excitation with different wavelengths.

Usually CDs synthesized via top-down method show size-dependent absorption properties [53, 91]. The surface functional groups also play a very important role in determining the absorption wavelengths. CDs synthesized by Peng *et al.* [92] showed a red shift in absorption spectra post functionalization with amino acids. Moreover, it is interesting to know that CDs prepared using the same precursor but different methodologies also show changes in absorption spectra. For example, CDs prepared using glucose as precursor via microwave-assisted, ultrasonic and hydrothermal or thermal procedures display different absorption peaks around 250–300 nm [93–95].

2.2.2 Photo-Induced Electron Transfer (PET) with CDs

Photo-induced electron transfer is an excited state electron transfer process. When a photon excites a molecule, an electron in a ground state orbital gets excited to a higher energy state (Figure 2.4). The excited electron is transferred from donor to electron acceptor. Such photoexcited molecule can act as an oxidizing agent or a reducing agent. In both oxidizing and reducing reactions an electron is delivered to an orbital that is higher in energy than where it previously resided. This is often described as a charge separated electron-hole pair when dealing

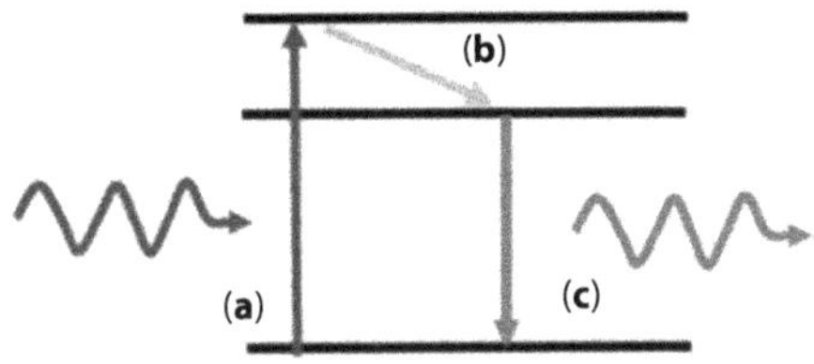

Figure 2.4 (a) Absorption of a photon causing an electron to move o an excited state, (b) the electron moves down in energy due to non-radiative processes, and (c) the electron recombines with the hole and gives off a photon.

with semiconductors. In the absence of a proper electron donor or acceptor it is possible for such molecules to undergo ordinary fluorescence emission. The electron transfer is one form of photoquenching. The photoluminescence in CDs (surface-passivated) can be efficiently quenched by electron acceptor or donor molecules in solution because photoexcited CDs are excellent electron donors as well as excellent electron acceptors. Due to this property, CDs offer the possibility of their potential use in light energy conversion and related applications.

2.2.3 Fluorescence/Photoluminescence (PL)

First, what is *luminescence?* Light is a form of energy that is expressed as either incandescence or luminescence. Luminescence is "cold light" that can be emitted at normal and lower temperatures. In luminescence, some energy source kicks an electron of an atom out of its lowest energy "ground" state into a higher energy "excited" state; then the electron returns the energy in the form of light so it can fall back to its "ground" state. With few exceptions, the excitation energy is always greater than the energy (wavelength, color) of the emitted light. There are several varieties of luminescence, each named according to the source of energy, or the trigger for the luminescence.

Fluorescence and PL are luminescence where the energy is supplied by electromagnetic radiation. Generally, PL is taken to mean "luminesce from any electromagnetic radiation," while fluorescence is often used only for luminescence caused by ultraviolet light, although it may also be used for other photoluminescence. Fluorescence is seen in fluorescent lights, amusement park and movie special effects, the redness of rubies in sunlight, "day-glow" or "neon" color, and in emission nebulae seen with telescopes in the night sky. Bleaches enhance their whitening power with a white fluorescent material.

Photoluminescence should not be confused with reflection, refraction, or scattering of light, which cause most of the colors we see in daylight or bright artificial lighting. PL generally produces light of a frequency that is lower than, but otherwise independent of, the frequency of the absorbed light. Their luminescence exhibits interesting wavelength dependence. That means that CDs show a tunable light-emitting range from blue to red under different exciting wavelength irradiations (Figure 2.5).

CDs show size-dependent PL and upconversion luminescence properties due to the multiphoton processes that lead to anti-Stokes type emission. The strong and tunable luminescence of carbon materials further enhances their versatile properties; in particular, the quantum effect in carbon is extremely

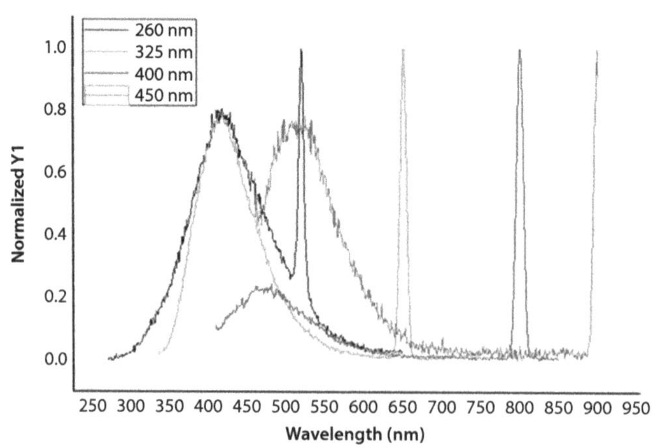

Figure 2.5 Photoluminescence of CDs.

important both fundamentally and technologically [96]. The optical behavior may reflect not only effects from particles of different sizes in the sample, but also a distribution of different emissive sites on each CD. Mechanistically, PL was attributed to the presence of surface energy traps, which become emissive upon surface passivation. They explained that there must be a quantum confinement effect of emissive energy traps on the surface in order for CDs to exhibit strong PL upon surface passivation [97].

So, the fluorescence, a type of PL, is one of the most important properties of CD. Fluorescence of the CD is an electronic phenomenon of molecules, which contain conjugated π-electrons. The Jablonski energy diagram (Figure 2.6) explains the principle of the fluorescence, i.e., the procedure of the electrons transition between different energy levels after absorbing certain light. When such molecules absorb light radiation, the π-electrons jump from the ground state (S_0, Figure 2.6) to higher energy state (S_1, Figure 2.6). To fluoresce, the majority of electrons go back from the higher vibrational levels to the lowest one and during this process energy is emitted, resulting in the emitted fluorescence. Due to the internal conversion, intersystem crossing and non-radiative relaxation, part of the energy is lost during electron transition, which explains the lower energy level of the emitted fluorescence compared to that of the absorbed light energy (named "Stoke's shift" or "red shift").

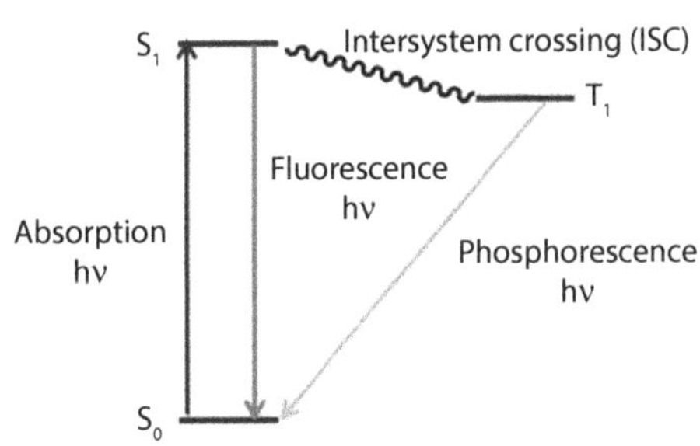

Figure 2.6 Jablonski energy diagram.

Though the PL of the CDs is one of the most fascinating features, it is not totally understood. The observed excitation wavelength-dependent emission of CDs may be the result of the optical selection of various nanoparticle sizes and/or the emissive traps on the surface of the CDs. The clear PL mechanism of CDs is not very clear and still open for debate. Since there are a variety of CDs, to explain the PL mechanism four PL principles have been confirmed:

1. The quantum confinement effect or conjugated π-domains determined by carbon core. The classic signature of quantum confinement is the observation of size-dependent optical absorption or PL. According to Li *et al.* [98, 29], PL of the CDs is a clear λex dependent of emission wavelength and intensity irrespective of different sized nanoparticles. There is dependency of the CD PL on excitation wavelength that is due solely to size differences rather than different emissive trap sites on similarly sized particles.
2. Mechanistically, Sun *et al.* [56] attributed the PL to the presence of surface energy traps, which become emissive upon surface passivation. The authors suggested that bright and colorful PL from the CDs might be attributed to the quantum confinement effect and presence of surface energy traps that became emissive upon stabilization as a result of surface passivation. So, mechanistically, the fluorescence emissions in CDs are attributed to radiative recombination of the carbon particle surface-trapped electrons and holes, where the large surface (relative to the particle volume) and diverse surface energy trapping sites in the small carbon nanoparticles are stabilized by the surface passivation agents.

3. Li *et al.* [29] reported that under strongly acidic or alkaline conditions, the PL is nearly completely quenched, whereas the CDs showed the strongest emission at pH 7.0.
4. The surface state determined by hybridization of the carbon backbone and connected chemical groups causes PL. Some organic molecules, e.g., diamine-terminated oligomeric poly(ethylene glycol) and poly(propionylethylene-imine-co-ethyleneimine), have no luminescent activity of their own, but when attached to the surface of C-dots, the CDs became luminescent (elaborated below in detail). As a result, it was believed that the surface of the CDs is the source of their luminescent abilities and that the surface passivation is the key to this ability. This may mean that the carbon surface, rather than the entire dot, is the essential part of CDs. The crosslink enhances emission (CEE) effect.

Thus, the origin of the PL of CDs has been assigned to several reasons in the literature which included optical selection of differently sized nanoparticles (quantum effect) (Figure 2.7), defects and surface states, surface groups, surface passivation, fluorophores with different degrees of π-conjugation, and the recombination of electron-hole pairs localized within small sp2 carbon clusters embedded within a sp3 matrix [99, 100].

Surface passivation is somewhat complicated, at least mechanistically, and a challenging thing to pinpoint and understand. Generally speaking, passivation changes the surface, such as the surface terminal atoms, becoming different after attachment of organic species [43, 78]. The unstable surface defects of CDs lead to reduction in radiative recombination, which gives rise to very low quantum yield in some CDs [101]. This limits the application of CDs in some specific fields. In order to overcome this problem, various physical and chemical characteristics

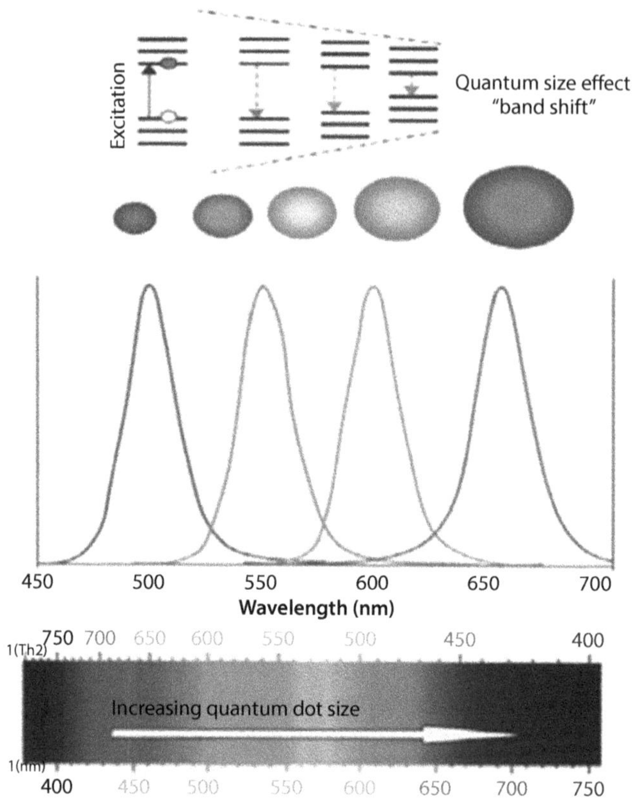

Figure 2.7 Quantum size effect and increase in quantum dot size. (Source: Physics Open Lab; Modern DIY Physics Laboratory for Science Enthusiast, English Post; Quantum Dots: A True "Particle in a Box" System)

have been adopted by researchers to enhance the luminescence. At the same time, passivation techniques are becoming very popular in increasing the quantum yield of CDs. Zheng *et al.* [33] have reported the synthesis of CDs with graphene oxide as precursor and NaBH4 as reducing agent. They have successfully enhanced the quantum yield from 2% to 24% by surface passivation. Wu *et al.* [11] have oxidized petroleum coke in a mixture of acids and further functionalized the CDs via hydrothermal ammonia treatment. This not only increased the quantum yield but also enhanced the fluorescence properties of CDs. In order to enhance the fluorescence and stabilize

CDs, doping with inorganic salts like ZnO, ZnS and different elements like N and S have also been attempted.

The QD of carbon (CDs) and previously discovered QD of silicon (Si-QDs) show that there may be more variety in how QD works. The mechanism for CD is different from that of Si-QDs, which are generally believed to be band-gap in origin. This means that for Si-QDs, the semiconducting nature of the material is important in how light is made, but for carbon something else is at work. To make CDs various methods were used, e.g., laser ablation of graphite. These particles formed aggregations and did not fluoresce. When the particles were treated with nitric acid, they did not fluoresce. The researchers then coated the particles with simple organic molecules, passivating the surface, which resulted in giving bright luminescence [101].

As mentioned above, the fluorescence of the QDs is a result of exciting the valence electron with a certain energy (or wavelength) and the emission of lower energy in the form of photons as the excited electron returns to the ground state combining with the hole. The fluorescence lifetime (the time the molecules remain in excited state before emitting a photon) is quite long, about 10–40 ns, which accounts for the stable and strong fluorescence. The larger the CD, the redder (lower energy) will be its fluorescence spectrum. Conversely, smaller dots emit bluer (higher energy) light. The coloration is directly related to the energy levels of the QDs. Quantitatively speaking, the band gap energy that determines the energy (and hence color) of the fluorescent light is inversely proportional to the size of the QD. Larger QDs absorb photons containing less energy, i.e., those closer to the red end of the spectrum.

Fluorescent intensity can be measured by the PL quantum yield of the QDs, which varies with the synthesis methods [102]. Brightness of a CD can be improved by adding another II-VI layer, such as ZnS. The same could be achieved by exchanging the capping ligands for primary amines [103]. The quantum yield of CD varies greatly with the synthesis method. CDs have been reported with quantum yield at 15% for green and 6% for red [104], 50% [103], all the way up to 80% [102]. Peng *et al.* mention that even a spectacularly low fluorescence of ~5% is sufficient for use in biological tests.

2.2.3.1 Multiphoton Excitation

When excited by UV and near-UV light, CDs show strong fluorescence. Moreover, CDs show luminescence with two-photon excitation in the NIR (800 nm) or long-wavelength light (500–1000 nm). This happens when multiple low-energy photons simultaneously arrive at fluorophores and interact with them so as to provoke excitation of an electron that is normally excited by one higher energy photon, after which multiphoton excitation is observed (Figure 2.8). It is this property of CD that makes it a very good agent for optical bioimaging [105]. Though there are many semiconductor QDs of heavy metals that exhibit multiphoton excitation, many in-vivo and in-vitro studies have shown that they are not biocompatible and have environmental and serious health concerns. Fortunately, CDs that are surface passivated not only exhibit bright photoluminescence in the visible region, their linkage with organic or other molecules via either covalent linkages or chemical adsorption make them biocompatible and two-photon luminescent microscopy imaging of the CDs that are internalized in cancer cells has been reported by many [106]. The first preliminary assessment of the bioimaging potential of CD was done using labeled *E. coli*

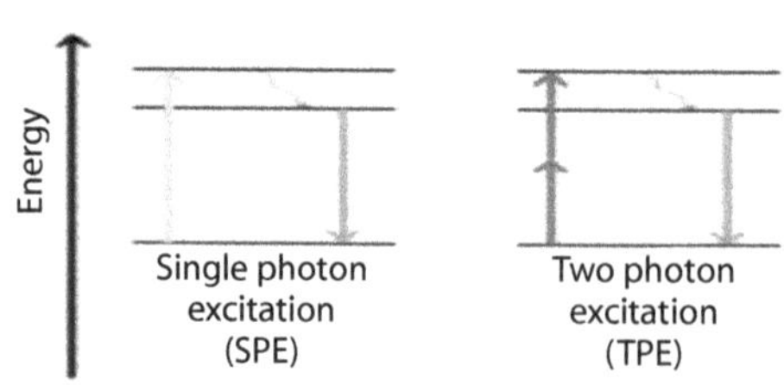

Figure 2.8 The electronic transitions observed in one-photon or two-photon excitation for multiphoton bioimaging.

ATCC 25922 cells with the CDs and cellular uptake of CDs by Caco-2 cells by confocal microscopy was observed [56].

2.2.3.2 Upconversion Photoluminescence

In upconversion one or more photons of lower energy are absorbed by a material, and then reemitted as a higher energy photon. Such materials are known as up-converters. CDs prepared by alkali-assisted electrochemical methods show not only size-dependent PL but are also excellent UPCL. Because of this attractive optical property CDs may be used for many promising applications. This property of CD may be attributed to the a) electron accepting and donor property of CDs, and also b) to multiphoton excitation.

The upconversion PL properties of CDs can be attributed to the multiphoton active process, in which the simultaneous absorption of two or more photons leads to the emission of light at a shorter wavelength than the excitation. Upon excitation in the NIR region, the PL spectra show a fixed emission peak and hardly shifts as the excitation wavelength varies. The emission spectra are nearly the same at each excitation wavelength, showing that the emission occurs from the lowest single state irrespective of the mode of excitation. Two-photon absorption (TPA) is the simultaneous absorption of two photons of identical or different frequencies in order to excite a molecule from one state (usually the ground state) to a higher energy electronic state. The energy difference between the involved lower and upper states of the molecule is equal to the sum of the energies of the two photons. Two-photon absorption is a third-order process, several orders of magnitude weaker than linear absorption. It differs from linear absorption in that the strength of absorption depends on the square of the light intensity, thus it is a nonlinear optical process. Two-photon absorption, phenomenon is also observed for the CDs, which emit visible light under excitation by near infrared or infrared light and hence can be a help in bioimaging [107]. They have synthesized CDs using ascorbic acid solution by a simple heating process at high temperature. The CDs solution was excited in the NIR region from 805 nm to 1035 nm, which emitted at 540 nm. The emission spectra were found to be stable even when the excitation wavelengths were changed. In another attempt, Zhang *et al.* [108] have synthesized nitrogen-doped CDs by one-step reaction between melamine and glycerol. The CDs displayed enhanced multiphoton upconversion fluorescence with a high two-photo absorption cross section, thus making it a potential agent for multiphoton single molecule bioimaging.

Upconversion fluorescence or photoluminescence is a very interesting and important property for such type of luminescent materials for their potential use in applications like biosensors (Figure 2.9). According to the frontier molecular orbital (FMO) theory, there exist two frontier orbitals, one highest occupied molecular orbital (HOMO) and another, lowest unoccupied molecular orbital (LUMO). To explain the differences of reactivity at each position in a molecule, the FMO theory was put forward by Fukui *et al.* in 1952[110]. Pi-electrons (frontier electrons as per FMO theory) are considered to be essential for the reactivity in photochemical processes because double or triple bond system takes part in the reaction. The FMO theory assumes that the highest occupied pi-orbital, i.e., HOMO of one molecule, interacts with unoccupied pi-orbitals, i.e., LUMO of the other molecule, as shown in Figure 2.10a.

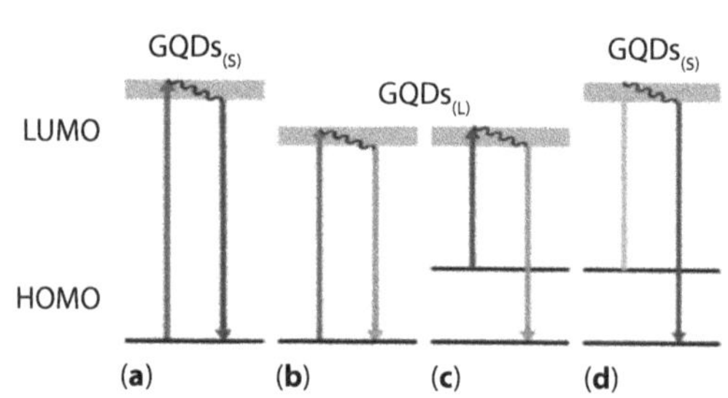

Figure 2.9 Upconversion photoluminescence by CDs. Facile preparation and upconversion luminescence of graphene QD. (Reproduced with permission from [109]; Copyright © 2012, Royal Society of Chemistry)

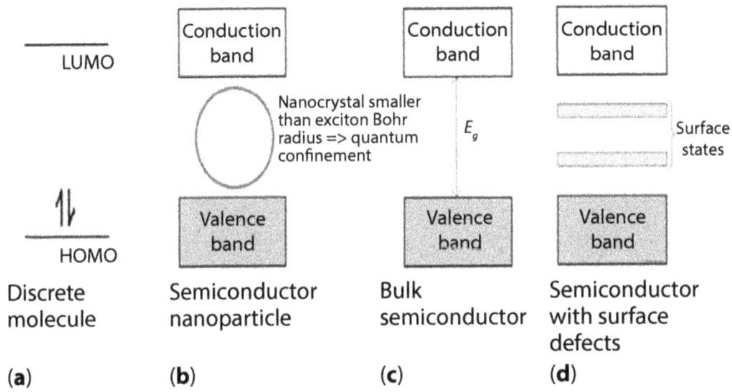

Figure 2.10 Electronic energy states of a semiconductor when it goes from (a) the single molecule to (c) the bulk material, passing through (b) the nanomaterial stage where quantum confinement is observed. (d) The effect of the existence of surface-defect states in the electronic energy state of the bulk material. (Adapted from [111])

2.2.3.3 *Lack of Blinking*

C-dots do not blink. Blinking is not intrinsically bad; but in some applications, the tendency of some QDs to blink on and off is a problem. It was speculated that no-blinking in CD is probably a result of the different luminescence mechanisms, which may be of advantage over traditional types of QDs for more possible applications. This property of CDs will no doubt lead to more research for application since these CDs have the potential to be used in medicine, whereas cadmium-based ones cannot [112]. CDs' different mechanisms of luminescence may already bring about an advantage over traditional types of QDs.

2.2.3.4 *Resistance to Photobleaching*

The fluorescence emitted by most of the fluorescent molecules gradually fades away with time or with activity. This phenomenon is known as photobleaching. It is a type of photochemical modification which leads to irreversible loss. In rare cases the fluorescent molecule can be switched on or reversed after partial emission [113]. A few fluorescent molecules, like semiconductor QD, display self-switching, which is also known as blinking of dots. In contrast to QD, CDs do not usually show a blinking nature and are found to be far more resistant to photobleaching that is found to display stable fluorescence over extended periods of time. Recently, Khan *et al.* [114] have displayed photoswitching of red emissive CDs, thus making them a potential candidate for high resolution bioimaging. The CDs were synthesized by microwave-assisted technique using chitosan as the carbon precursor. The process was followed by surface passivation using poly ethylene glycol. Large numbers of cycles were studied and this was probably the very first report on photoswitching of CDs using visible light by alternating excitation at 401 and 639 nm laser (Figure 2.11). The experiments confirm the involvement of the electron transfer process, which is proposed as a mechanism for photoswitching. According to the researchers, a loss of electron disrupts the emission process, which can be repaired by high-energy laser. It was estimated that new sites were not created by photoactivation but the same present emissive sites are reactivated.

Wang *et al.* [113] also studied in detail the possible causes of photobleaching in CDs. The researchers have stated that

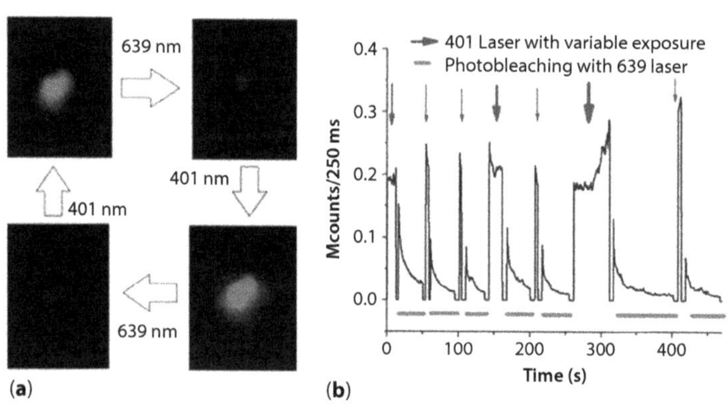

Figure 2.11 (a) The on-off switching observed by alternating 401 nm and 639 nm laser excitation. (b) Photon counts in a hybrid photomultiplier detector show photo decay and subsequent gain in intensity (photoswitch) using a 401 laser. The blue arrow shows the 401 nm pulses (detected in channel 1) and the red bars show the photo decay with a 639 nm laser (detected in channel 2). The 4th and 6th cycles, with a slightly higher exposure time of blue light, result in a better recovery of fluorescence. (Reproduced with permission from [114])

the extent of photobleaching mainly depends on the duration and intensity of exposure to light. It was also observed that the extent of photobleaching is less in a highly concentrated CD solution. The influence of oxygen was also investigated, which showed that if oxygen dissolved in the CD solution is removed by nitrogen purging, fluorescence decay still takes place, but the rate is greatly reduced. Other factors which contribute to photobleaching include formation of free radicals and change in surface functional groups. Hence, it was concluded that intense and highly stable fluorescence of CD is possible with well-dispersed nanoparticles and surface passivation. All of these factors have to be taken into consideration for making the best of the present available CDs and make them ideal candidates for various advanced bioimaging applications.

2.2.4 Photocatalytic Property

A good photocatalyst should be able to utilize visible and/or near UV light and should be inexpensive and eco-friendly. To design photocatalyst the most important properties are its band gap energy, its chemical composition and surface modifications. Size-controlled CDs showed tunable emission in between near-infrared UV wavelength, making them a promising performer as photocatalyst.

Very pure, high quality, crystalline CDs having excellent aqueous dispersibility, and remarkable down- and up-converted PL properties, have been demonstrated to possess high photocatalytic activity under visible irradiation, and potential for high-efficiency complex catalyst design. Li *et al.* [115] has demonstrated that their strong and tunable photoluminescence enables the CDs to efficiently utilize the full spectrum of sunlight. Interestingly, the PL from CDs can be quenched efficiently by either electron acceptor or electron donor molecules in solution, indicating that photoexcited CDs are excellent electron donors and electron acceptors [116]. CDs have found important and wide applications in energy and catalysis. In converted

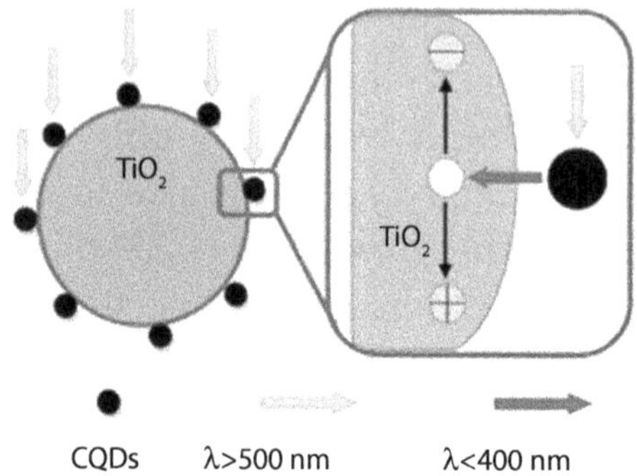

Figure 2.12 Possible catalytic mechanism for TiO_2/CDs under visible light. (Reproduced with permission from [118])

PL, CDs were shown to possess excellent up-converted PL (UCPL), which enables the design of high performance, complex catalyst systems based on CDs for efficient utilization of the full spectrum of sunlight [117].

Li *et al.* [118] have reported electrochemical synthesis of CDs which possess size-dependent PL spectra and excellent upconversion fluorescence. These enhanced fluorescence properties have been utilized to design photocatalyst systems consisting of TiO_2/CDs and SiO_2/CD complex (Figure 2.12). After illumination of the complex photocatalyst, CDs absorb visible light and emit at shorter wavelength, which in turn excites TiO_2/SiO_2 to generate electron-hole pairs. These electron-hole pairs react with adsorbed oxidants/reducers to produce active oxygen radicals which further cause degradation of the dyes [119, 120]. Based on the same possible mechanism, scientists have published various other research of excellent CD-based photocatalytic processes such as CDs/ZnO, CDs/Fe_2O_3, CDs/Cu_2O, etc. [121, 122, 29].

2.3 Chemically Inert

Chemically inert materials have much less or no ability to react with other elements, molecules or compounds. Chemical inertness depends mainly on the valency of elements. Elements usually react with other substances because they have to satisfy their valency by filling their octets, but those elements which already have a completely filled octet show no or much less reactivity towards other elements and rarely react with other elements. This is known as chemical inertness. CDs are chemically inert, hence are stable, due to which they can be used in various applications.

2.4 Easy Functionalization

Functionalizing the surface of CDs can be done through surface chemistry such as covalent binding, co-ordination and π-π interactions. The majority of CDs are rich in oxygen-containing groups, which makes them feasible to form covalent binding. The presence of functional groups makes CDs soluble in water, ethanol, acetone and DMSO. This property leads to easy surface passivation, which is a very important step in stabilization of CDs. It is also believed that surface passivation improves the fluorescence intensities. So far, various molecules, such as polymers, have been used for surface modification of CDs [123-125]. Some reports have also proved that bare CDs or naked CDs display feeble emission as well as low quantum yields [126, 127, 43].

2.5 Water Solubility

The property of CDs that is very suitable and attracts its use in biosystems is its aqueous solubility. This property of CD is due to a) inherently low and amphiphilic nature depending on the surface capping materials [43], b) presence of large amount of carboxylic and hydroxyl groups at their surfaces, thus endowing them with excellent water solubility and suitability for subsequent functionalization with various organic, polymeric, inorganic or biological species. Carboxyl moieties on the CD are also suitable for chemical modification and surface passivation, which enhances both fluorescence properties and physical properties.

2.6 Low Toxicity

Cell cytotoxicity can be explained as the ability of some chemicals or mediator cells to destroy living cells. The healthy living cells undergo the process of necrosis or apoptosis post reaction with the cytotoxic compound. Cytotoxicity of any compound or nanoparticles plays a very important role for development of therapeutics as well as theranostic processes. CDs are majorly made of carbon element and hence show less cytotoxicity that has higher biocompatibility as compared to heavy metallic semiconductor QD as well as other metallic nanoparticles. Recently, various research groups have evaluated the in-vitro cytotoxicity as well as in-vivo cytotoxicity of CDs. Wang *et al.* have carried out toxicity studies of CDs synthesized using different combinations of precursors and molecules of surface functionalization [128]. CDs and free PEG 1500N were not found to be cytotoxic when tested on MCF-7 and HT-29 cells. The results also showed that CDs were not toxic below the threshold concentrations of PPEI-EI and PEI. But free poly acrylic acid (PAA), as well as CDs functionalized with PAA, were found to be harmful to cells at relatively low concentrations. Liu *et al.* also reported similar results and combining the two it can be predicted that the cytotoxicity of CDs depends on the surface passivation agent to a great extent [116]. For in-vivo cytotoxicity, CDs were synthesized using raw soot and refluxing them with nitric acid [129]. Surface functionalized (PEG2000) CDs were intravenously given to mice. After 14 days various tests were carried out on the mice. Genotoxicity results displayed normal values for micronucleus cell frequency of hematopoietic cells in bone marrow and polychromatic erythrocyte/

normochromaic erythrocyte ratio. The blood analysis, histopathology and body parameters were within normal ranges. Hence, CDs were found to be nontoxic to mice at tested doses and can serve as a potential molecule to be used for fluorescence in in-vivo imaging, tumor cell tracking and other related applications. Sharon's group [130, 131, 50, 72] have reported a cytotoxicity study of various CDs synthesized using natural and chemical precursors on different cells using MTT assay. They found that the effects of CDs on various cell lines were negligible and a small amount of cells were found dead due to the alkalinity of CDs' solution.

2.7 Biocompatibility

Carbon dots passivated by various molecules (such as PEG, PEI, NaOH) have been evaluated for their cytotoxicity and *in-vivo* toxicity and biocompatibility for different cell lines such as MDCK epithelial kidney Vero cell lines [130], L929 Vero cell line [106] and compared with other toxic QDs such as CdSe/ZnS, etc. Their successful use in drug delivery, bioimaging and uptake by various cell lines has suggested that the carbon dots are biocompatible, and their performance as fluorescence imaging agents is competitive. The CDs are widely accepted as being more advantageous over conventional organic dyes, genetically engineered fluorescent proteins, because of their resistance to metabolic degradation [132]. CDs may be biocompatible and bioinert due to their solubility in water as well as their robust chemical inertness. Recently it has been shown that CDs are not only biocompatible but can also cross the blood-brain barrier [133].

2.8 Summary

This chapter has given a brief introduction to various properties of CD that has made it a suitable material for application in biosystem. Optical properties, such as photoluminescence, multiphoton excitation, upconversion photoluminescence, lack of blinking, resistance to photobleaching and its photocatalytic property, have been found to have a main role in bioimaging. Whereas, its chemical inertness, easy functionalization, water solubility, low toxicity and biocompatibility have found favor in its use in drug delivery.

3

Carbon Dots and Conjugates

Nanotechnology is truly a portal opening on a new world.
Rita Colwell, Director, National Science Foundation

3.1 Introduction

As a scientist, when we consider the word *conjugates* or *conjugated system*, we often get confused between *conjugate* and *composite*. According to the *Encyclopaedia Britannica*, a conjugate can be defined in three possible ways. The simplest is:

i. A conjugate refers to "a compound formed by the joining of two or more chemical compounds."
ii. Whereas, it is chemically defined by the "Brønsted-Lowry theory of acids and bases," where the term conjugate refers to "an acid and base that differ

from each other by a proton and when both (acid and base) react, the acid forms its conjugate base while the base forms its conjugate acid," i.e.:

$$\text{Acid} + \text{Base} \rightleftarrows \text{Conjugate Base} + \text{Conjugate Acid}$$

The acid loses a proton to become its conjugate base A^- as the base B accepts a proton to become its conjugate acid HB^+.

iii. According to third definition, "Conjugation is the overlap of p-orbitals across a σ bond (sigma bond). In transition metals, d-orbitals may overlap." The orbitals have delocalized electrons when there are alternating single and multiple bonds in a molecule. Bonds alternate in a chain so long as each atom has an available p-orbital. Conjugation tends to lower the energy of the molecule and increase its stability.

Thus an accepted definition of conjugated system is a system of p-orbitals connected with delocalized electrons in molecules with alternating single and multiple covalent bonds that generally lower the overall energy of the molecule and increase stability. Lone pairs, radicals or carbenium ions are examples of conjugated system. A conjugated compound may be linear, cyclic, acyclic or mixed. When two groups, e.g., C:O a carbonyl-group and OH a hydroxyl group (separately in a molecule, each has distinctive properties), are conjugated together they form the carboxyl group (COOH), which has an entirely different property. Graphene, graphite, conductive polymers and carbon nanotubes (CNT) are conjugated system. Some examples of conjugated systems are given in Figure 3.1.

Composite material or *Composites*, as they are generally called, is synthesized using two or more constituents having different chemical and physical properties; when they are combined together they form a product having different properties.

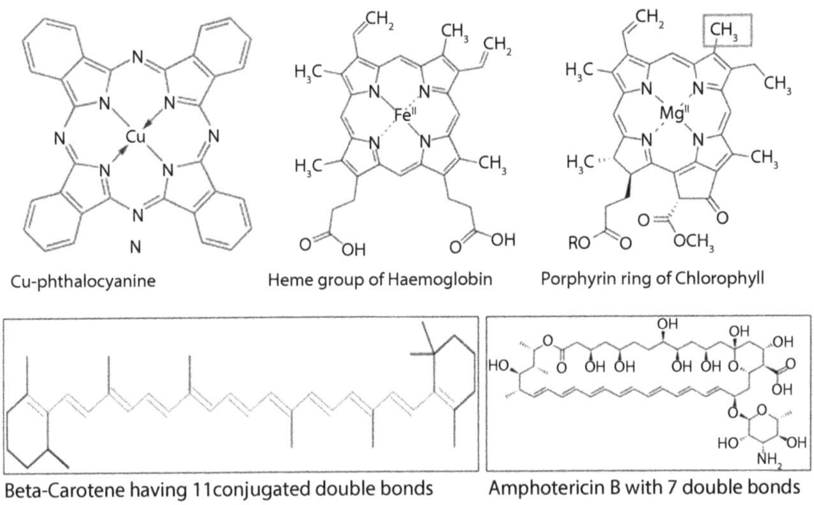

Figure 3.1 Examples of conjugated systems.

However, unlike conjugates, each of them retains their separate and distinct characteristic within the composite. Fiber-reinforced polymer is an example of composite material that is produced for reasons such as getting a stronger, lighter, or less expensive material when compared to traditional materials. Recently, polymer composites are being used for sensing, actuation, computation, communication, aircrafts, spacecraft and as robotic materials. Common composites that are used in our everyday life are mortars, concrete, metal composites, ceramic composites and reinforced plastics. These composites have found great applications in infrastructures such as buildings, bridges, boat hulls, swimming pool panels, race car bodies, shower stalls, bathtubs, storage tanks, imitation granite and cultured marble, etc. Composites are made of two or more different materials; whereas in nanocomposites there is a solid matrix or polymer in which nano-size fillers, such as carbon nanotubes, are embedded.

In this chapter we will be discussing conjugation and composite formation at nanoscale level between carbon dots (CDs) and other materials.

3.2 Why Conjugation of Carbon Dots?

In order for any nano or quantum size particle to be suitable for its application as a theranostic agent it needs: Desired particle size and size distribution – as they affect the *in-vivo* distribution, stability, drug loading and drug release ability, uptake by cells, biological fate, toxicity, and targeting ability of these delivery systems, availability to a range of cellular and intracellular targets, easy mobility and capacity to cross the blood-brain barrier following the opening of endothelium tight junctions by hyper-osmotic mannitol, to treat neurogenerative diseases and brain tumors [134]. Smaller particles have larger surface area-to-volume ratio, hence can attach more drugs and, being near to the particle surface, release drug faster [135]. Thus, control of particle size provides a means of tuning drug release rates.

Another important criterion is that the surface charge property of a nanosized particle is characterized by the zeta potential. According to Chowdhary *et al.* [136], zeta potential values indicate positive charge on the surface of CDs. Since CDs have above ±30 mV zeta potential they are stable in suspension, because the surface charge prevents its aggregation. With its help it can be determined whether CD is encapsulated within the center of the conjugate or conjugate is attached to the surface of CDs.

Due to the specific photoluminescent (PL) property of CDs, its water solubility and biocompatibility, it has found applications in many fields. Such applications have demanded fabrication of CD conjugates. We often use the word functionalization or surface modification when CD is conjugated with a functional moiety. Functionalization promotes the application of CD, for example, in bioimaging. Functionalization of CDs with PPEI–EI and PGE1500N has successfully resulted in its application in chemical sensing by Sun *et al.* [56], Cao *et al.* [24] and Yang *et al.* [137]. Conjugation of CDs with another moiety helps in selectively recognizing a target and giving sensitive signal response. For various biomedical applications, synthesis of CDs and polymer has been explored [138]. When CDs are conjugated with amino and carboxylic group they exhibit more PL and are made water soluble [28].

Due to their biocompatibility and ease of encapsulation, CDs and polymer composite are widely accepted as excellent prolonged drug delivery carriers. Natural polymers like chitosan, alginate, polypeptides, PLGA, PEG, etc., have been found very suitable for making polymeric CD composites. Due to its unique polymeric cationic character, chitosan has been extensively examined in the pharmaceutical industry for its potential use in the development of controlled drug delivery and enhanced drug payloads.

Doping of CDs is also a form of conjugation. According to Chandra *et al.* [139], sulphur-doped CDs composited with phenyl-C60-butyric acid methyl ester (PCBM), exhibits wide band gap of 4.43 eV with a high open circuit voltage (V_{OC}) of 617 mV along with very high (37%) fill factor (FF) and high fluorescence quantum yield of 11.8% without any additional surface passivation. Here, PCBM acts as the electron transporting layer. This CD composite will be useful in the fabrication of solar cells and unique fluorescent probe for further applications. Polymeric conjugates are also needed for drug delivery. Covalent conjugation of CDs with dyes increases their photophysical properties [140]. With an increase in polymeric therapeutics, polymeric CD composites are finding their uses in drug delivery.

Functionalization of CDs or surface modification of CDs with different chemicals is done for applications in different biological, medical and engineering areas. Morphology and structure of nanoparticles, as well as the spatial location and adsorption state of nanoparticles, are critical in achieving optical functionalization. The tunable surfaces of nanoparticles provide versatile scaffolds for numerous applications in biomedicine. Both natural and synthetic chemical can be used for functionalization. Functionalized CDs can impart enhanced cellular internalization ability, non-cytotoxicity, and improved payload

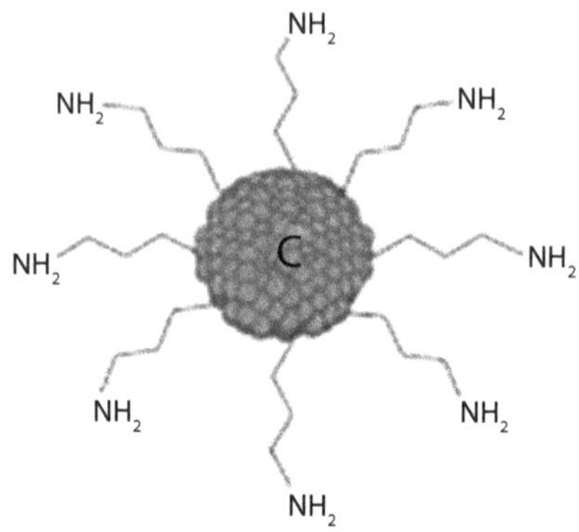

Figure 3.2 NH_2 functionalized carbon dot; here, NH_2 can act as a linker.

binding capacity necessary for effective intracellular delivery [141]. Surface functionality provides selective or specific recognition required for biosensing [142, 143]). On the basis of the surface functionalities of CDs, they can be used as an efficient tool for diagnosis and treatment of diseases. Moreover, the flexibility of functionalization of CDs with various groups, such as $P_{25}TiO_2$, makes them able to absorb different wavelengths and this improves the capacity.

The functionalization molecules act as linker for attaching other molecules to the CDs (Figure 3.2). This is very important for drug delivery. CDs can act as an excellent dual-functional nanocarrier for delivering therapeutic payload. The drug release profile can be altered and/or adjusted using polymer coating and pH responsive linkers.

3.3 Types of Carbon Dot Conjugates and Their Applications

The common conjugates that are comprehended to be attached to CDs are: *Biogenic compounds* (such as proteins/peptides, DNA, RNase, lipids, Fe-amino-clay, dihydroxylipoic acid, folic acid protoporphyrin); *inorganic nanomaterials* (gold, silica, ZnO, Mn^2 + ZnS, Fe_2O_3 and Europium); *organic materials* (covalently conjugated dyes and polymers); *anticancer drugs; anti-neurodegenerative drugs; antibiotics* and *MWCNT*. These conjugates are prepared for the purpose of delivering drugs to the desired site and/or tracking them. Whatever may be the mechanism of drug delivery, ways have to be devised of linking drugs to carriers such that release of the drug takes place at the right place and at the right time. Drug carrier linkages may be covalent or noncovalent and conjugates based on thiols, hydrophobic interactions, electrostatic interactions, and avidin-biotin complexes have been investigated. Drug carriers, whether actively or passively targeted, are generally directly conjugated to the carried drugs or encapsulate them in some form. Encapsulation can be manifested in many ways, including enclosure in a polymer matrix, encapsulation within a dendrimer or within a vesicle. Encapsulating drug carriers potentially have the advantage of allowing controlled drug release at the target tissue, yet completely shielding the drug until it is released. The use of cleavable covalent bonds is perhaps the most promising strategy. For instance, specific short peptide sequences have been developed that are enzymatically cleaved near the target. But cleavage can also take place non-enzymatically: most common are pH triggered linkers, including hydrazones, cis-aconityl linkers, electron-rich trityl linkers, acetals, imines, cysteamine hydrochloride and others

3.3.1 Biogenic Compounds Conjugated with Carbon Dots

Biogenic compounds are produced due to metabolic activity of living organisms. They are necessary for the living processes. These compounds may be organic (with at least one C-H bond, i.e., carbohydrates, lipids and proteins) or inorganic (with no C-H bond, i.e., skeletons, shells, silica compounds, etc.) in nature. They may be either constituents, extracts or secretions of living beings such as microbes, fungi, algae, plants or animals. Many times these compounds get transformed and stored by undergoing changes into other compounds such as chalk, limestone (secretions of marine animal shells), wood, pearls, silk, ambergris, etc. The actively metabolizing biogenic materials include, proteins, peptides, DNA, RNA, lipids etc. Here, in this chapter, we are concentrating on metabolically active biogenic compounds that have been found to successfully form conjugate with CDs and that can be used for drug delivery. Proteins and peptides are some such compounds.

3.3.1.1 CDs Conjugated with Proteins/Peptides

Conjugation of peptides and proteins to CDs can be achieved by i) Using EDC, 1-ethyl-3-(3-dimethylaminopropyl) carbodiimide, condensation to react carboxyl groups on the CD surface to amines; ii) Directly binding protein or peptide to the CD surface using thiolated peptides or poly-histidine residues; or iii) Adsorption or noncovalent self-assembly using engineered proteins.

The specific fluorescence emission bands and broad overlapping excitation spectra of multiple color dots property of CDs have been utilized for a variety of *in-vitro* and *in-vivo* applications, such as fluorescent biological labels, for multiphoton fluorescent imaging; *in-vivo* imaging; labeling of cancer marker [104, 23, 144, 145]; and for intracellular drug (Streptavidin) delivery to mammalian cells, especially by conjugating it with cationic peptides [146], i.e., a nine-residue biotinylated L-arginine peptide. This short arginine-rich peptide sequence is also known as protein transduction domains (PTDs), which includes part of the segments from the human immunodeficiency virus 1 (HIV-1) transcriptional activator Tat protein, the *Drosophila* homeotic transcription protein antennapedia (Antp), and the herpes simplex virus structural protein VP22, and homopolymers of arginine attached to CDs. A successful attachment of Streptavidin was achieved by linking the C-terminus of the peptide (biotin-C6-(L-Arg)9) peptides to

CDs via Streptavidin-biotin linkages by Lagerholm *et al.* [146]. It helped in the intracellular uptake of CDs in many mammalian cell types such as embryonic mouse fibroblasts (Swiss 3T3), human endothelial cells (HeLa), and human osteoblast-like cells (MG63). The CDs were found to accumulate in intracellular vesicles characteristic of endosomes and lysosomes.

Lu *et al.* [147] have demonstrated that CDs are strongly two-photon active with pulsed laser excitation in the near-IR which exhibits bright luminescence in the visible. According to them, the two-photon absorption cross sections of the CDs are comparable to those of the best-performing semiconductor QD or core-shell nanoparticles. Moreover, they demonstrated the potential of CDs for cell imaging in human breast cancer MCF-7 cells, which became brightly illuminated when imaged on the fluorescence microscope with excitation by 800 nm laser pulses. The CDs labeled both the cell membrane and the cytoplasm of MCF-7 cells. However, it did not reach the nucleus. But later, the accumulation of CDs in the cell as well as the nucleus was demonstrated by Santra *et al.* [148] and Stroh *et al.* [149] by conjugating CDs with membrane translocation peptides TAT (a human immunodeficiency virus-derived protein). TAT facilitated the translocation of the tissue by overcoming the cellular membrane barrier and enhancing the intracellular labeling efficiency.

Real-time molecular tracking in live cells is done by fluorescent tags that are bright, photostable, biocompatible, and of molecular size so that it can minimize physical hindrance. For this purpose, oxygenated graphene QD (GQD) that is very similar to CDs was conjugated by Zheng *et al.* [150] with protein (insulin) without impairing the functionalities of the protein or largely altering the size, weight, and charge state. Other proteins that were tried are neuropeptide Y, bovine serum albumin, immunoglobulin G, concanavalin A and nerve growth factor. Insulin conjugated GQD revealed the real-time dynamics, i.e., distribution, internalization, and recycling, of insulin receptors in adipocytes for the first time.

3.3.1.2 CD Conjugates of Amino: Carboxylic Acid Ratio

It has been observed by Hu *et al.* [151] that when CDs are synthesized from citric acid (CA) and 1,2-ethylenediamine (EDA) by microwave-assisted pyrolysis, along with controlled pH and reaction time one of the important parameters was mole ratio of amine (NH_2) to carboxylic acid (COOH) moieties of the precursors of the CDs, for increased synthesis of CDs. They explained that the negatively charged CDs are obtained only when the mole ratio of NH_2/COOH is 0.25–0.80; whereas, if by contrast, when the quantity (mole) of NH_2 is equal to or larger than that of COOH in citric acid, only positively charged and neutral CD species are formed, which are predominantly covered by the surface-attached ammonium and amido moieties. They used capillary zone electrophoresis (CZE) coupled with UV absorption and laser-induced fluorescence for detection of the complexity of CDs.

3.3.1.3 CDs Conjugated with DNA

To check the interaction between CDs when conjugated with single-stranded DNA (ssDNA) and double-stranded DNA (dsDNA) and their impact on the fluorescent behavior of CDs, Milosavljevic *et al.* [152] used CDs synthesized from the citric acid coated with polyethylene glycol (PEG). They found that the interaction of ssDNA with CDs did not influence the fluorescence intensity of CDs; whereas with dsDNA, the fluorescence intensity of CDs increased significantly. Their results provide a basic understanding of the interaction between CDs and DNA. The benefit of this result can be utilized in biosensing.

There are many applications for DNA-conjugated CD such as for drug delivery, for detection of antigen, DNA and micro-RNA, for sensing and bioimaging. The advantage of using CDs-DNA conjugate is that it functions as a fluorescent vehicle to deliver drugs. Ding *et al.* [153] synthesized DNA-CDs. They used genomic DNA isolated from *Escherichia coli* for this purpose. During DNA–C=do synthesis, ribose ring was opened and thus collapsed, nitrogen was released, and several new bonds (C–OH, N–O, and N–P) were formed. However, the other covalent bonds of DNA remained. This offered the presence of many chemical groups, such as amino or hydroxyl groups on DNA-CDs, that were bound to facilitate further functionalization. These DNA-CDs were highly biocompatible. They have envisaged its use as a new fluorescent vehicle for cell imaging as well as for drug delivery.

Another application of CDs conjugated with DNA is creating a hybrid hydrogel for sustained release of drugs [154]. For this purpose, amine functionalized CDs were conjugated to 5′-phosphate termini of cytosine (C)-rich ssDNA by phosphoramidate linkage. The attached drug was Doxorubicin. The whole conjugate was enclosed in hydrogel that assisted sustained release of the Doxorubicin. Here, CD acts as a cross-linker for network formation. Moreover, the presence of CD enables tracking of hydrogel dissolution and drug cargo loading in hydrogel. The *in-vitro* time and pH-dependent release profile of the drug from hydrogel was studied, which showed that hydrogel was stable for a month at normal physiological pH, whereas complete dissolution and sustained release of the drug molecules were achieved over 10–11 days in acidic pH (the tumor microenvironment is known to have acidic pH). Slow killing of HeLa cells was also achieved using this CD/DNA/Doxorubicin encapsulated in hydrogel.

One of the most widely used applications of CDs is in bioimaging and sensing; CDs + DNA conjugates can sense DNA-specific biomolecules that can be observed by their fluorescence. For example, Xu's group [155] designed a target-induced assembly of a sandwich nanostructure for CDs and aptamer-based fluorescent detection of thrombin. Thrombin is an enzyme present in blood plasma that clots the blood by converting fibrinogen to fibrin. *Aptamers* are single-stranded DNA or RNA (ssDNA or ssRNA) molecules. *Aptamers* assume a variety of shapes due to their tendency to form helices and single-stranded loops. Hence, they are extremely versatile and bind targets with high selectivity and specificity. They attached two thrombin aptamers on the surfaces of CDs and silica nanoparticles to produce two separate assembling units; and have shown that in the presence of thrombin, CDs got attached around the surface of silica nanoparticles due to the specific aptamer–thrombin interaction. The thus-formed sandwich structure was quantitatively detected by fluorescence emission spectra and fluorescence microscopy images, which helped in quantifying the thrombin.

Another application of DNA-conjugated CD was demonstrated for gene therapy. A lot of attention is being paid to gene therapy these days because it can treat many diseases. DNA-conjugated CDs have been found to be a suitable carrier for safe and efficient delivery of therapeutic payload into cells.

Wang *et al.* [99] used Alkyl-PEI2k CDs for surface passivation and gene delivery. After passivation with Alkyl-PEI2k, the zeta potential of CDs became 17.33 ± 1.97 mV, while the naked CDs bearing negative charges had a zeta potential of −20 mV, confirming that there was presence of amino group of the Alkyl-PEI2k on the surface of CDs. Other properties that they found very useful were that the nanovector possesses good stability, monodispersity with narrow size distribution, high biocompatibility, low cytotoxicity, high gene delivery efficiency with the necessary feature of visualizing the delivery and photoluminescence. Moreover, they found that Alkyl-PEI2k CDs were suitable for (pDNA) for l gene therapy and optical molecular imaging. According to Wang *et al.* [99], positive charges are the most suitable surface charges for nanoparticles to be used in gene delivery. The gene binding was attributed to the electrostatic interactions between the amine groups of the polymer and phosphate groups of genes, which eventually help form polyelectrolyte complexes. Therefore, they proposed that the positively charged surface of the Alkyl-PEI2k CD is suitable for binding to a negatively charged siRNA. siRNA is a short interfering RNA, which is used as a RNA interference (RNAi) tool for inducing short-term silencing of protein-coding genes. siRNA is a synthetic RNA duplex designed to specifically target a particular mRNA for degradation. When they added heparin to this CD complex, the siRNA was released from the complexes.

Also, CD is perceived by Kim *et al.* [156] to be suitable material to monitor the association/dissociation of polymeric carrier/plasmid DNA (pDNA) complex during transfection (*transfection* is the process of introducing naked or purified nucleic acids into eukaryotic cells). To use CD as efficient gene delivery vector, they quenched the fluorescence of CD by Au nanoparticles and then the surface of CD and Au was modified with cationic polymer, polyethylenimine (PEI), and finally treated with non-labeled pDNA. It was found that CD-PEI/Au-PEI/pDNA ternary nano-assembly is a highly efficient hybrid transfecting agent with high cell viability under the optimum conditions and it was very effective at monitoring the carrier/pDNA dissociation in a non-labeled manner. Therefore, it can be said that it provides an efficient strategy to study the mechanistic aspect of polymer-mediated pDNA delivery. After extensive study, Kim *et al.* [156] found that a ternary complex consisting of PEI-functionalized CD (PEI-CDs), PEI-functionalized gold colloids (PEI-Au), and plasmid DNA (pDNA) could be used for *in-vitro* DNA transfection and dynamic monitoring of plasmid cellular trafficking. They exhibited the utility of CD to mediate *in-vivo* transfection. According to them:

> "pDNA-facilitated complexation with two carrier systems, namely fluorescent nanoparticle and the quencher nanoparticles, not only plays a crucial role in imparting excellent physicochemical properties to the polyplex deemed ideal for efficient cellular uptake, but also confers unique optical properties which could provide a highly intriguing method to monitor the complexation and especially the dissociation of polyplex inside the cytoplasm after cellular uptake."

3.3.1.4 CDs Conjugated with RNase and SiRNA

One of the challenges of using CD for molecular imaging and *in-vivo* molecular tracking is to enhance the fluorescence intensity of CD. A new way of synthesizing CD with high quantum yields (24.20%) was demonstrated by Liu *et al.* [157] by using multifunctional ribonuclease A (RNase A) as a biomolecular templating agent (i.e., as an assistant reagent) (RNase A@CDs) under microwave irradiation. By adjusting microwave reaction time and power, the fluorescent color of the RNase A@CD can be adjusted. When the excitation wavelength is altered from 300 to 500 nm, the photoluminescence (PL) peak of RNase A@ CD exhibits redshifts from 450 to 550 nm and at 380 nm maximum intensity is noted, i.e., its Stokes shift is about 80 nm. The PL intensity gradually decreases with the increase in pH reaching maximum at a pH 2 and the emission peaks also show a redshift, which Liu *et al.* credited to the high activity and perfective dispersion of RNase A in a lower pH solution because in high pH solution, RNase A tends to form RNase A warped CD nanoclusters.

The cell imaging, 3D confocal imaging and TEM showed that the RNase A@CD enters into the cytoplasm through endocytosis and finally enters into the nucleus. The RNase A@CD inhibits the growth of MGC-803 cells. By injecting RNase A@ CD into the tumor they were able to do the *in-vivo* imaging of gastric cancer cells. The RNase A@CD is a potential conjugate that can be used as theranostic agent (*theranostics* is an integrated therapeutic and diagnostic system that can simultaneously monitor the real-time response of therapy).

Wang *et al.* [158] have also shown the utility of CD as an efficient siRNA nanocarrier for therapy of gastric cancer cells. They used CD synthesized from citric acid by microwave-assisted method and passivated it with nitrogen using tryptophan. This CD was highly water soluble, biocompatible and had high quantum yield. It was further conjugated with the PEI (polyethylenimine) (CD@PEI) and used to deliver Survivin siRNA into human gastric cancer cell line MGC-803. The siRNA induced an efficient knockdown for Survivin protein to 6.1% and arrested the cell cycle progression in G1 phase as well as cell caused apoptosis, thus confirming the potential of the CD@PEI conjugate as both imaging agents and as siRNA nanocarriers for Survivin siRNA delivery. And the results indicate that these nanocarriers can be useful in a broad range of siRNA delivery systems for cancer therapy.

A multi-functionalized CD conjugated with folate and reducible polyethylenimine passivated CD (fc-rPEI CD) can also be used as theranostic nanoagent for gene delivery in lung cancer therapy [159]. In lung cancer therapy, folate is conjugated as targeting ligand because of folate receptors overexpression on the surface of cancer cells. This conjugate emits visible blue photoluminescence and excitation at under 360 nm and can encapsulate multiple siRNAs (EGFR and Cyclin B1). EGFR is an epidermal growth factor receptor that is overexpressed in lung cancer and also other carcinoma; it is a transmembrane receptor tyrosine kinase protein. And Cyclin B1 is a regulatory protein involved in mitosis; it begins to increase during G2, peaks in mitosis, and is rapidly degraded before the cell cycle is completed. The fc-rPEI-CD is highly biocompatible and a good siRNA gene delivery carrier for targeted lung cancer treatment. The positive surface charge on CDs can complex with negatively charged siRNAs. The fc-rPEI-CD could function as a siRNA carrier and release siRNA in a reducing environment. fc-rPEI-CD/pooled siRNAs selectively accumulate in lung cancer cells through receptor-mediated endocytosis, resulting in better gene silencing and anticancer effect. No need to mention again that CDs are promising bioimaging agents because of their persistent photoluminescence. Moreover, gene silencing therapy knockdown oncogenetic signaling pathway and induce apoptosis through specific interaction of siRNA and corresponding mRNA offers a lot in therapeutic segment, and this specificity can promote treatment efficiency by targeting personalized genetic defects in the tumor.

3.3.1.5 *CDs Conjugated with Lipid*

Conjugation of CDs with lipids is mostly done by encapsulating the CDs inside the lipid micelle. However, covalent attachment of CD to phospholipids has also been explored as a possibility [160] to be used as vehicle for studying lipid-bilayer dynamics, i.e., the dynamic properties of cell membranes. Moreover, the interactions of proteins, drugs, and viruses with the cell surface can also be understood by this conjugate. Nandi *et al.* [161] have shown that the CD-modified phospholipids get readily incorporated within biomimetic membranes (including solid-supported bilayers), vesicles, and into actual cellular membranes. Using a CD-phospholipid probe the effects of polymyxin-B (a cytolytic peptide), valproic acid (a lipophilic drug), and amyloid-beta (a peptide associated with Alzheimer's disease) upon bilayer fluidity and lipid dynamics have been studied.

3.3.1.6 *CDs Conjugated with Folic Acid*

Generally, folic acid conjugation with not only CD but even with other nanoparticles is done to use it as a navigational molecule due to its high expression in most cancer cells. Sharon's group has used folic acid-conjugated CD (synthesized from sorbitol) for delivery of therapeutic payload (Doxorubicin and anticancer drug) to desired target [50]. The surfaces of the CDs were protected with bovine serum albumin (BSA). BSA makes the conjugate more biocompatible and allows attachment of a high amount (~86%) of drug (Doxorubicin). The complex was internalized in HeLa cells where it followed first order release kinetics of Doxorubicin at physiological as well as slightly acidic pH. Moreover, the killing rate of cancer cells was more when Dox@CD was used than when free CD was used. FA-mediated targeting killed less normal cells.

A unique method of synthesis of uniformly distributed CD is done by the acid treatment of folic acid with H_3PO_4 [162] by hydrothermal procedure. The advantage of using folic acid as precursor of CD was that it provided an inherent doped effect in nitrogen composition and its size was ~ 13 nm. The surface of this CD was coated with poly(amidoamine) (PAMAM-NH_2) dendrimer. The size of PAMAM-NH_2 coated CD was ~30 nm. This nanoconjugate was successfully used as fluorescent sensor for Pt(IV) ion in aqueous media (in the form of chloroplatinate ion). The limit of detection of this conjugate was demonstrated to be 657 nM with an accuracy of the method of 0.13% (as RSD, n = 10) and sensitivity of 78 nM. The dissolved platinum ions can be quantified in the range of 6–96 μM with an accuracy of 2.5%. This sensor worked on quenching mechanism with a good accuracy (RSD = 0.13%) and was moderately interference free.

Campos *et al.* [162] performed an acid treatment of folic acid at 100 °C to get CD. Whereas in 2016 Bhunia *et al.* [52] successfully synthesized CD from folic acid by treating folic acid with alkali NaOH, which is known as passivating agent in CD synthesis (Figure 3.3).

Moreover, CD synthesized from folic acid by Bhunia *et al.* was found to be nontoxic and selectively entered into cells that overexpressed folate receptors. Due to the fluorescent nature of CD, such cells could be distinguished by fluorescent microscopy. So, it can be said that this CD is suitable for selective imaging of cancer cells that express the folate receptor on their surface.

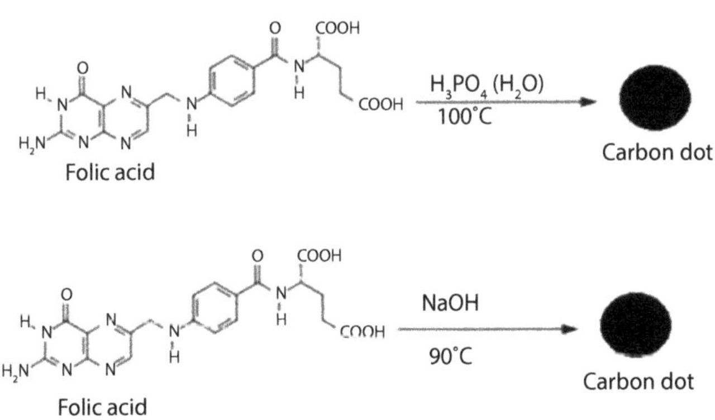

Figure 3.3 Carbon dots can be synthesized from Folic acid as precursor under acidic ($H_3PO_4(H_2O)$) or alkaline (NaOH) condition.

3.3.1.7 CDs Conjugated with Chitosan

Chitosan is a natural polymer (a linear polysaccharide), which is obtained from the alkaline deacetylation of chitin, obtained from shells of shrimp and other crustaceans. It is composed of randomly distributed β-(1→4)-linked D-glucosamine that is the *de-acetylated* unit. It is a biodegradable and biocompatible material, widely used in synthesizing hydrogel, for the adsorption of drugs and their controlled release. It is used for oral as well as subcutaneous delivery of drugs for treating cancer and diabetes. It is also used as precursor for synthesis of CD [44, 163, 164, 80]. Even though the precursor is the same, the PL properties of such CD varies with

i. pH; for example, CDs that were prepared at pH 3 exhibited sharp PL peak whereas at pH 1 or 5 a broad emission peak was obtained in the PL spectra [165].
ii. Passivating agents; for example, CDs passivated with PEI have a quantum yield of 13.15% and PL decay lifetime is 6.183 ns. Whereas, CD s passivated with PEG have a quantum yield of 7.01% [166] and PL decay lifetime is 4.825 ns.

Moreover, CD conjugated with PEI is better bioimaging agent as compared to CDs conjugated with PEG. This suggests that selection of the right polymer groups for the surface is very important for bioimaging. Chitosan-derived CD, when conjugated with low molecular weight gelators, has been found to be a very fruitful application for ion sensing in aqueous medium [167] as a sensor for detecting Pb^{2+}, Cr^{6+}, Cu^{2+}, Fe^{3+}, Pb^{2+} and Mn^{2+} [168] and also anion F^-.

PEG-chitosan@CD hybrid nanogels exhibit enhanced gelation properties and increased PL; hence, they have many applications. Wang *et al.* [169] used them in two-photon fluorescence (TPF) bioimaging, pH and near-infrared (NIR) light dual-responsive drug release, and synergistic therapy.

The biogenic precursors of CD have great impact on the properties of CDs. CDs prepared from tea and then conjugated with chitosan yield stable, soft but tough, chitosan-carbon dots [170] nanocomposite hydrogel films that are strong, thermo-stable and have wettability of hydrogels.

Chitosan is not only used for the adsorption and controlled release of drugs systems but also has been tried as scaffold for cell and tissue growth. Very recently, Campos *et al.* [171] fabricated thermoresponsive microgels-based encapsulated CDs synthesized from D-lactose by hydrothermal method and coated with PEG. These particles were ~4 nm spherical in shape and monodispersed, which had covalently attached PEG. The thermoresponsive material to which it was introduced was pNIPAM microgels. Incorporation of CDs in pNIPAM-CDs@PEG hybrid microgel system is confirmed by TEM, fluorescence microscopy, energy-dispersed X-ray (EDX) spectroscopy and dynamic light scattering (DLS).

Chitosan is also used as precursor for synthesis of CDs by simple carbonization, which offers some extra advantages such as simultaneous conjugation with amino group on its surface. These CDs exhibit bright luminescence with absolute quantum yield (QY) of 4.34%, are excitation- and pH-dependent and have upconversion fluorescence behavior and have potential applications in biosensing, cellular imaging and drug delivery [40].

3.3.1.8 CDs Conjugated with Digitonin

Digitonin (DG)-conjugated CD is used for detection of atherosclerosis (caused by increase in blood cholesterol level leading to cholesterol deposits or plaque formation in cardiac blood vessels, which is the main cause of heart attack or myocardial infarction). This probe can selectively bind to a cholesterol-rich area in a solution that can be imaged by fluorescent microscopy and IVIS (In Vivo Imaging System) Spectrum images [172, 173].

3.3.2 Inorganic Heteroatoms Conjugated with CDs

Initially, as-prepared CDs were used directly without any further modifications. Recently, several nanocomposites based on CDs are reported using various inorganic materials that have organic photovoltaic properties for photocatalytic activities for CO_2 conversion or the degradation of organic dyes [174, 175, 37, 176].

CDs have been conjugated/doped with many metal (Cu, Zn, Gd, Au, Ag) and non-metal (Si, B, P,N, S) heteroatoms (atoms apart from carbon or hydrogen are also called heteroatoms); because they enhance the properties of CDs (such as increased quantum yield in far red and near IR, decreased photodamage) and have found application in biosensing, bioimaging, and theranostics. They have been extensively reviewed by Xu *et al.* [177].

Imaging of living tissue up to about one mm in depth can be done by two-photon excitation microscopy. In two-photon microscopy the wavelengths of the two exciting photons are longer than the wavelength of the resulting emitted light. Hence, it is different from fluorescence microscopy where the excitation wavelength is shorter than the emission wavelength. For two-photon excitation microscopy, near-IR excitation light is used that can also excite fluorescent dyes. The advantage of using IR is that for each excitation, two photons of IR light are absorbed, which minimizes scattering in the tissue and the background signal is strongly suppressed. Because of these two effects, the penetration depth of these microscopes is increased. Thus, due to its deeper tissue penetration, two-photon excitation microscopy has efficient light detection and comparatively much less phototoxicity.

Conjugation of CDs with inorganic-organic nanosystems has found applications in two-photon imaging and biosensing of pH variations. Kong *et al.* [178] have fabricated a highly sensitive and selective CDs-based two-photon fluorescent probe that can monitor 6.0–8.5 pH gradients and can be used in 2-photon imaging. It can biosense living cells and tissues at a depth of 65–185 μm with good cell permeability. This CDs-based probe is highly photostable and has low cytotoxicity. It has performed real-time imaging and biosensing of pH in living human lung cancer A549 cells, mouse LLC-MK2 cells, and tumor tissues (> 100 μm depth). It is believed that it can be further developed for *in-vivo* determination of metal ions, proteins, DNA, and other biologically relevant species.

3.3.2.1 CDs Conjugated with Gold Nanoparticles

After the milestone work of Tian's group in 2010, which exploited amplification of Raman signal by using Au nanoparticles with silica or alumina shells and evolved a new SERS operation mode, called shell-isolated nanoparticle-enhanced Raman spectroscopy (SHINERS), Luo *et al.* [179] synthesized Au@CD. The Au NP was synthesized by reducing $HAuCl_4$ and CD was prepared by electrochemical method. They modulated the average diameter of Au@CD by adjusting the ratio of $HAuCl_4$ to CDs used for fabrication. This conjugate has an ultrathin CDs shell of ca. 2 nm. They used a suspension of Au@CD having an average diameter of ca. 24 nm as a substrate for surface-enhanced Raman scattering (SERS). This Au@CD conjugate exhibited a higher SERS effect for Rhodamine 6G (Rh6G) as compared to the suspension of pure Au nanoparticles of approximately the same size. The enhanced SERS effect of Au@CD, along with the enhanced chemical interaction with aromatic probe molecule, was attributed to the fact that CDs have a special affinity for aromatic molecules. Therefore, they can be used in SERS detection of aromatic molecules, especially toxic melamine or explosive nitroaromatic compounds.

Some of the anatomical as well as physiological constraints for drug delivery to cancer cells by conventional medicine are poor distribution of anti-cancer drug to tumors as well as body fluid, non-specific interaction between drug and other cells, resistance to drug due to repeated administration, etc. Hence, efforts are underway to fabricate nano-size targeted drug delivery vehicles fabricated using nanoparticles that can enhance the permeation and retention (EPR) effect through the capillaries of solid tumors. Gold nanoparticles of various shapes and sizes have been found by Sharon's group to be a suitable metal for this purpose, e.g., gold nanorods, gold nanotriangles and gold nanospheres by Pandey *et al.* [131]; Dharmatti *et al.* [180] and Pandey *et al.* [181] respectively.

Gold nanorod (GNR) conjugated with CD is found especially useful for photothermal therapy as well as delivery of therapeutic payloads because of its typical size and shape related features. Sharon's group has shown that conjugation of CDs with GNR and Doxorubicin can act as a tri-pronged molecular vehicle for targeted (folic acid was used as synaphic targeting molecule) drug delivery, photothermal therapy (as it shows rapid burst of drug release under the influence of near infrared [NIR] radiation 808 nm) and biological imaging. In this conjugate the CDs attached to the surface of GNR acted as the protection for the GNRs' surface. Moreover, CDs reduced the toxicity of the nanorods due to the presence of the cationic surfactant CTAB, and drug loading capacity of CDs@GNRs was as high as ~94%.

3.3.2.2 CDs Conjugated with Silica

Conjugate of microparticles of mesoporous silica oxide and CDs (mesoSiO_2/CDs) have found profound application in bioimaging. Conjugates of mesoSiO_2 with CDs have been done by:

i. Separately preparing CDs using the electrochemical method and opaque nanospheres of meso-SiO_2 using tetraethyl orthosilicate (TEOS) SiO_2 and then conjugating them [182, 183]. These two particles were conjugated by a facile hydrolytic process to fabricate CDs/SiO_2 porous nanocomposites (Figure 3.4). When this conjugate was used as a catalyst for catalytic selective oxidation of *cis*-cyclooctene, and it showed high catalytic

Figure 3.4 SEM image of CDs/SiO_2 nanocomposites. Insert is the corresponding SEM image of a typical single particle. (Adapted with permission from [183])

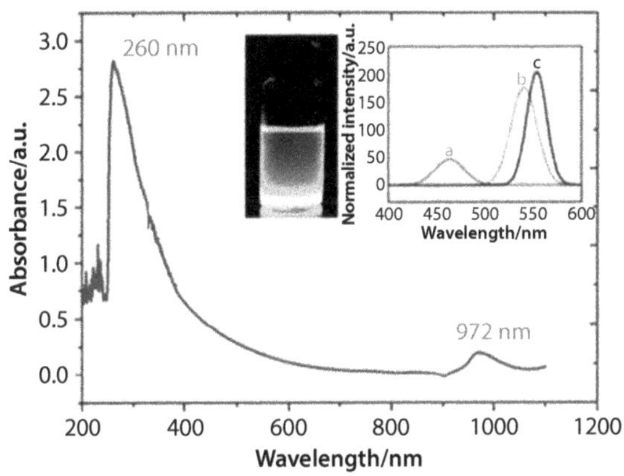

Figure 3.5 UV-Vis spectra of meso-SiO_2/CD synthesized using rice husk. Inset shows photoluminescence spectra excited at (a) 250, (b) 350, and (c) 450 nm. (Adapted with permission from [184])

activity and stability for the selective oxidation of *cis*-cyclooctene under visible light irradiation, using tert-butyl hydroperoxide (TBHP) as a radical initiator and oxygen from the air as an oxidant at 80 °C.

ii. By using rice husk (as it has both silica and carbon) as precursor to get biogenic meso-SiO_2/CDs using a novel hot injection method commonly used for

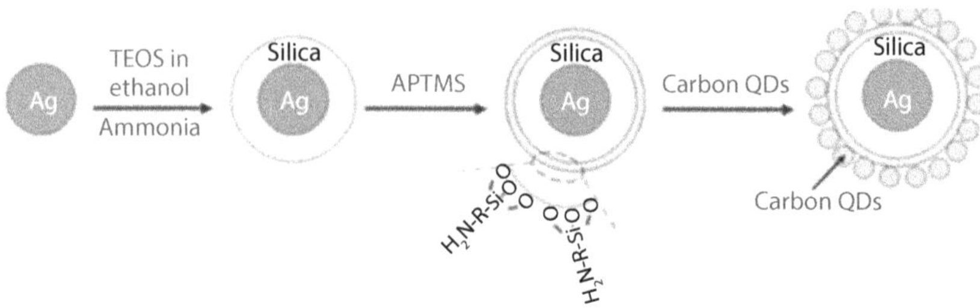

Figure 3.6 Layer-by-layer assembled silver/CD plasmonic luminescent nanoparticles. (Adapted from [185])

semiconductor QD (Figure 3.5) [184]. This conjugate was found useful in labeling yeast cells. When tested against Vero cells, meso-SiO_2/CDs were found to be highly biocompatible.

There is a unique attempt to prepare a complex of Silver/CD plasmonic luminescent nanoparticles with silica interlayer through layer-by-layer assembly (Figure 3.6). The thickness of silica layer can be used to tune the plasmonic and luminescent property. This is because due to silica layer the position of the plasmon peak gets red shifted, which weakens the plasmon intensity. And the luminescence intensity gets enhanced by an increase in the thickness of the silica interlayer [185].

In another attempt [186] to enhance the fluorescence of CDs, it was adsorbed onto Ag@SiO_2 nanoconjugate (Figure 3.7). Here, instead of silica layer, as mentioned above, silica-shell is used to control the distance between silver-core and fluorescent CDs; which resulted in a four-fold increase in fluorescence intensity of CDs. Here, the silica-spacer thickness was adjusted to optimize the metal fluorophores distance. Moreover, it also exhibited enhanced upconversion.

3.3.2.3 CDs Conjugated with ZnO

The conduction band of zinc oxide (ZnO) is –4.05 eV and valence band is –7.25 eV (vs. vacuum) [187], while the work

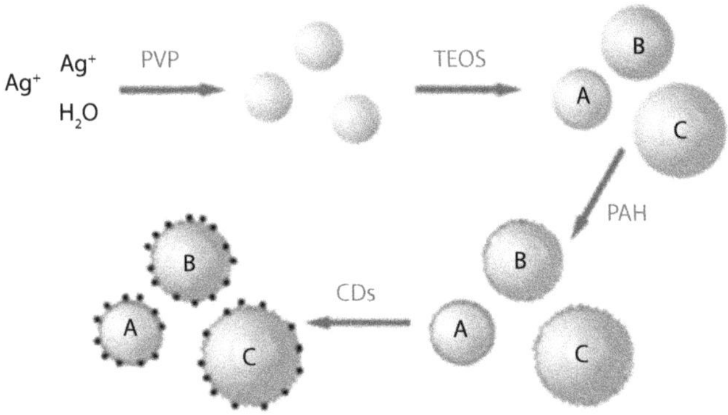

Figure 3.7 Schematic representation of the preparation of the CDs-adsorbed Ag@SiO2 composite nanoparticles. (Adapted with permission from [186])

function of CDs is −4.5 eV [188]. The electron-hole pairs generated in ZnO under UV light can photoinduce electrons transfer from ZnO conduction band to CDs, which will separate the photogenerated electron-hole pairs and hinder the charge recombination. Therefore, Yin *et al.* [189] utilized CDs to prepare ZnO/CD to get improved charge separation and the reduction of charge recombination; hence, enhanced *photocatalytic* efficiency. They prepared ZnO/CD conjugate by coating the layers of CDs onto the surface of ZnO by a sol-gel method combined with a spin-coating process. This conjugate has been found to have photocatalytic property as investigated by using Rhodamine B. Yin *et al.* have suggested that the photocatalytic property is enhanced because there is some kind of electronic interaction between ZnO and CDs, which contributes to both the improvement of charge separation and the reduction of charge recombination. The layers of CDs coating on the surface of ZnO layer decides the degree of enhancement of photocatalytic activity. Four layers of CDs showed the highest photocatalytic activity.

3.3.2.4 CDs Conjugated with CdS

For enhancing the *photocatalytic* activity of CDs-based conjugate, CD-doped CdS (CD/CdS) microspheres have also been tried [190] using Rhodamine B as a test dye under solar-simulated light irradiation. To hydrothermally prepare self-assembled flower-shaped CdS microspheres, CD was doped, as it is an electron acceptor. This conjugate CD/CdS exhibited higher photocatalytic activity and higher photodegradation rate of Rhodamine B than single-phase CdS. Liu *et al.* have attributed this enhancement to the electrons trapped by the CDs and the hindrance to the recombination of photogenerated electron and hole pairs. Moreover, the upconverted PL property of CD supported the utilization of visible light more efficiently.

Recently, several CD-based photocatalysts have been achieved, e.g., CDs embedded with mesoporous hematite [191], a quasi-cube hematite/CD [122] that could efficiently degrade benzene and methanol, CD/Ag_3PO_4 as well as CD/Ag/Ag_3PO_4 photocatalysts [51], and CDs-decorated $SrTiO_3$ that could convert near IR photon energy to photocurrent [192], which also yielded high electron-hole separation rate of the heterojunction between $SrTiO_3$ and CDs. Since CDs can significantly enhance the carrier density and annihilation of fluorescence it is incorporated into different micro- or nanoparticles to make an efficient catalyst. The CDs in the composites can efficiently trap electrons, thus hindering the recombination of photogenerated electrons and holes.

3.3.2.5 CDs Conjugated with Strontium Oxide

Many algae are known to metabolize production of high-energy molecules such as fatty acids and triglycerides. Nowadays, efforts are underway to use these high-energy molecules as feedstock for biodiesel [193].

A conjugate of nanoparticles of SrO with CDs is used as *catalyst* for fatty acid methyl esters production using *Chlorella vulgaris* as feedstock [194]. For this conjugate, first CD is synthesized by microwaving poly ethylene glycol (PEG 400). CD is then sonochemically adhered to the surface of $Sr(NO_3)_2$ (precursor for SrO nanoparticles) and finally calcinated for 4 hours in inert atmosphere at 650 °C to get SrO-CDs nanoconjugate.

SrO-CDs nanoconjugate was used to catalyze the transesterification of the lipids in the alga *Chlorella vulgaris* with methanol. It afforded a lipid conversion value of 97 wt% after 2.5 min of microwave irradiation, which was better than the catalytic activity of commercial SrO, which yielded 45.5 wt% after 2.5 min of microwave irradiation.

3.3.2.6 CDs Conjugated with Gadolinium(III)

Gadolinium(III)-doped CDs of uniform ultrafine (3–4 nm) particle size has been fabricated as *dual fluorescence MRI probes* for biomedical applications (first reported by Bourlinos *et al.* [96]). This conjugate is monodispersed, it forms stable dispersions in water and exhibits bright fluorescence, strong T1-weighted MRI contrast and low cytotoxicity. Synthesis of this composite was done by thermal oxidation of an organic salt and a gadolinium (Gd)(III) complex. The Gd-doped CDs could be used for multimodal imaging. Gd(III)-doped CD has been found to be comparable to commercial Gadovist. According to Rehor *et al.* [195], "incorporation of the Gd compound into solid matrices brings extra advantages to T1 contrast enhancement in comparison to a 'free' T1 contrast agent." Moreover, earlier various inorganic Gd compounds, such as $Gd_2O(CO_3)_2$ and Gd_2O_3, have shown good T1 contrast enhancement [196].

3.3.2.7 CDs Conjugated with Europium

Europium (Eu^{3+}) is a rare earth ion. It can coordinate with carboxylate group and link to another neighboring carboxylate group [197]. Moreover, Sastry *et al.* [198] have shown that Eu^{3+} has higher affinity to oxygen-donor atoms that originates from phosphates than that which originates from carboxylate groups. Hence, Zhao *et al.* [199] decided to fabricate an off–on fluorescence platform by utilizing the differential property of affinity to oxygen donors originating from different sources, i.e., between phosphate (Pi) and carboxylate. They designed a europium-adjusted off–on fluorescent probe with carboxylate functionalized CDs (Figure 3.8) to develop a simple method of phosphate

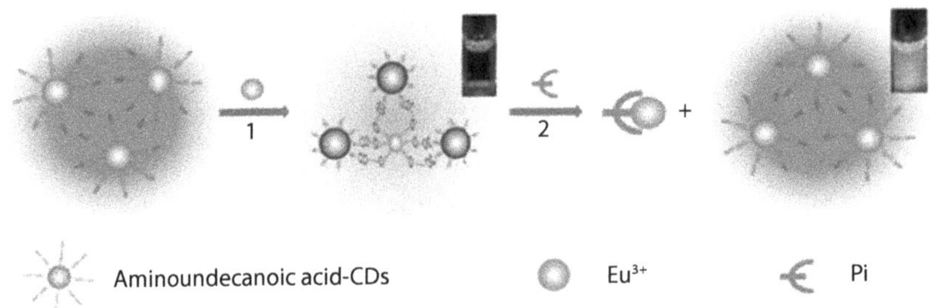

Figure 3.8 Schematic representation of Pi detection based on the off–on fluorescence probe of CDs adjusted by Eu^{3+}. (Adapted with permission from [199])

(Pi) detection even in a complex matrix. CDs were synthesized by condensation reaction between citric acid and 11-aminoundecanoic acid. The CDs thus formed had a lot of carboxylate attached on the surface. This probe, apart from being a simple and cost-effective method for determination, is a very sensitive probe.

Also, CDs conjugated with Europium have been used [40] for the detection of tetracycline. Tetracycline is an antibiotic used to treat a number of infections. However, it has damaging effects like renal toxicity and hemolytic anemia. If tetracycline is released in the environment or water it is expected to induce bacterial drug resistance. Therefore, a sensitive and selective detection is developed by fabricating Europium-doped CD (Eu-CDs). This conjugate exhibits blue photoluminescence, which is quenched by tetracycline; the quenching is due to a strong inner filter effect mechanism between Eu-CDs and tetracycline. The detection limit of Eu-CDs is 0.3 μM at 465 nm.

CDs@Eu-based nanoscale coordination polymers (Eu-NCPs) have been used [200] for detection of 2,6-pyridinedicarboxylic acid (DPA). DPA is a biological warfare agent. The presence of many hydroxyl and carboxyl groups on CDs helps as a starting point for self-assembly of Europium (Eu^{3+}) and 5′-guanosine monophosphate disodium (GMP) on the surface of CDs, which forms CDs@Eu/GMP nanocomposites/conjugate and acts as nanosensor. The Eu/GMP NCPs serve as the DPA recognition sites. When they sense DPA, this sensor exhibits the strong emission of CDs (remains constant as a reference) and weak emission of Eu^{3+} (this gets enhanced when it binds to DPA). Thus, the strong and weak fluorescence (of CDs and Eu^{3+} respectively) ratio displays ratiometric fluorescent detection of DPA with very good selectivity. The ratio of F615/F469 linearly increases with the increase in DPA concentration (in the range of 25 nM–5.0 μM) with a detection limit of 5.1 nM. This nanosensor can also detect DPA in the serum samples.

3.3.2.8 CDs Conjugated/Doped with Nitrogen, Sulphur, Phosphorus and Boron

For traceable drug delivery system, CDs doped with nitrogen has been found to be a good material [201]. Moreover, unlike metal heteroatoms, non-metal heteroatoms like nitrogen and sulphur are nontoxic and biocompatible. Nitrogen can be doped onto CDs either alone or along with other non-metal dopants such as sulphur. N-doped CDs are highly fluorescent because nitrogen atom injects electrons into CDs and changes the internal electronic environment, thus increasing their fluorescence properties and quantum yield. N-doped CDs can be prepared biogenically from dried shrimp using a facile hydrothermal approach that yields CDs having multifunctional groups, i.e., amino, carboxylic acid and hydroxyl. It was used for targeted and traceable delivery of boldine to human breast cancer cells (MCF-7 cells) and SH-SY5Y (human neuroblastoma). The boldine was loaded on the surfaces of the CDs via hydrogen bonding and the release was triggered by acidification of the intracellular vesicles in which the carriers were located. Since the N-doped CD was functionalized by multifunctional groups it had very good stability. Boldine was easily internalized by endosomes and then released into the nucleus causing death of the MCF-7 cells. Such biocompatible N-doped CDs hold potential for drug delivery with reduced side effects caused by a heavy dose of chemotherapy. Similar strategies of preparing N- doped CDs were followed by Dong *et al.* [202], who hydrothermally prepared it from soymilk; Huang *et al.* [203] did one-pot hydrothermal synthesis from *Bombyx morisilik.*

Nitrogen and sulphur co-doped carbon dots (N, S-CDs) synthesized from heparin sodium that exhibited strong blue-green luminescence with good biocompatibility, low toxicity and high cell viability have also been used as optical imaging probes for the detection of Fe^{3+} [204] with high sensitivity, because Fe_3 can strongly quench the fluorescence ofN, S-CDs.

N-doped thiol functionalized PEGylated CDs have found application for fluorogenic detection of Hg^{2+} because they are highly sensitive (picomolar level) and have high selectivity over other metal ions,. This strategy is expected to offer a new low-cost and sensitive sensor in biological and environmental applications.

According to Jana *et al.* [205], since boron and nitrogen are positioned on the left and right side of Carbon in the periodic table, doping or co-doping with both these two neighbor heteroatoms to CD has special fundamental and practical significance for carbon-based nanomaterial. Co-doping of CDs with both boron and nitrogen have yielded very high fluorescence [29].

Not much attention has been paid to S-doped CDs. For S-doping, suitable compounds that are a source of sulfur (S) are sulfuric acid, waste frying oil, sodium citrate as C source along with sodium thiosulphate as S source, etc., as reported by Zhu *et al.* [67], Li *et al.* [29] and Hu *et al.* [100] respectively.

Doping CDs with phosphorus (P) has been successfully tried because doping CDs with P atoms is expected to alter the optical and electronic properties [206–209].

The heteroatom conjugated CDs have found potential for application in biosensing and bioimaging.

3.3.3 Carbon Dots Conjugated with Organic Material

Though CDs are intrinsically not cytotoxic, there are some indications that due to the presence of some of the passivating molecules on the surface they do exhibit cytotoxicity [210, 211]. Murali *et al.* [212] have reported that PEG is very sensitive to sonolytic degradation and PEG degradation products can be toxic to mammalian cells. Therefore, there have been attempts to synthesize polymeric CDs. Polymers that have found application in medicinal science have been investigated for conjugation with CDs. Single-chain, branched, and hyperbranched polymers have been tried for conjugation with CDs for various applications, some of which are described below.

3.3.3.1 PEG (Polyethylene Glycol)

Polyethylene glycol (PEG) is a polymer of ethylene oxide having a molecular mass below 20,000 g/mol. It is also known as Polyethylene oxide (PEO) if the molecular mass is above 20,000 g/mol, or Polyoxyethylene (POE), which is a polymer of any molecular mass. PEG can be branched or hyperbranched. Common medicinal uses of PEG include as a base in laxatives and an excipient in many pharmaceutical products. When attached to proteins for a therapeutic purpose, PEG ensures its slow clearance from the blood; it can also be used as coating of vector (virus) in gene therapy. Studies conducted by Yang *et al.* [101] on several human cancer cell lines and by Zhao *et al.* [213] on human kidney cell lines and by many others [214] have shown that PEG 1500N; O,O0-bis(3-aminopropyl) polyethylene glycol, PPEI-EI, poly(propionylethyleneimine-co-ethyleneimine), and PEI, polyethyleneimine functionalized CDs are nontoxic [94]. One of the advantages of conjugating CD with PEG is to make CDs nontoxic to human cell lines. Another advantage of PEG-conjugated CDs is that one can attach nuclear localization signal (NLS) peptide to it for cell nucleus targeting and imaging; as demonstrated by Yang *et al.* [215] using MCF7 and A549 cells, which presented a possibility that the PEG-conjugated CDs may serve as cell nucleus imaging probes.

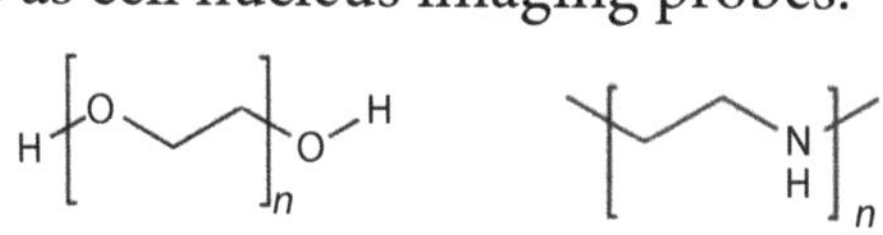

PEG $H–(O–CH_2–CH_2)_n–OH$ PEI $(C_2H_5N)_n$

3.3.3.2 CDs Conjugated with PEI (Polyethylenimin) or Polyaziridine

Polyethylenimin (PEI) is a negatively charged cationic polymer. It is used in many detergents, adhesives, water treatment agents and cosmetics. It has limited medicinal uses because it is toxic at higher concentrations, though it is used in cell culture as an attachment promoter of weakly anchored cells and as a transfection agent [216]. One of the most important uses of PET is its application in space craft to capture and remove CO_2. And in electronics as low-work function modifier.

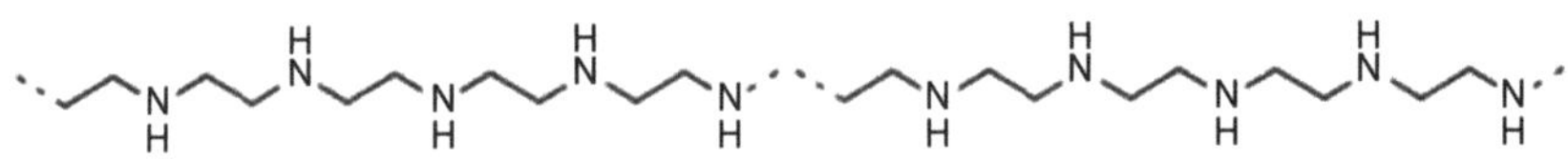

Linear PEI (contains all secondary amines);

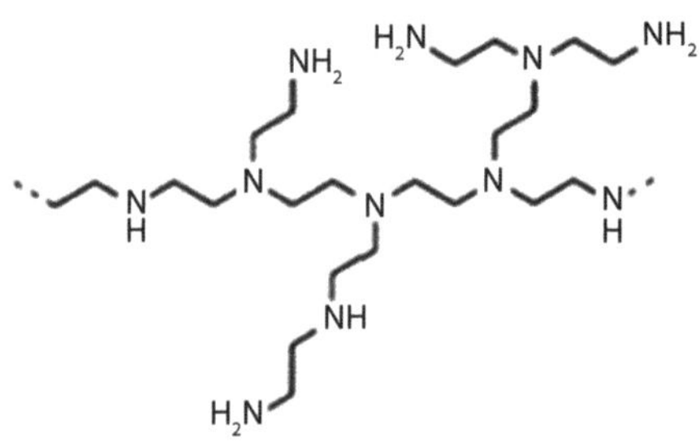

Branched PEI (contains primary, secondary and tertiary amines);

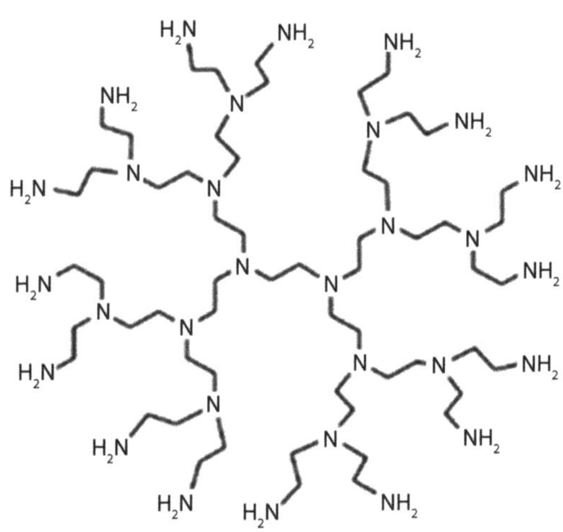

PEI-generated dendrimers are derived from its polycationic character.

High fluorescence quantum yield (42.5%) CDs functionalized with polyamine can be easily synthesized by one-step pyrolysis using citric acid as precursor along with branched PEI at a low temperature (< 200 °C). The product displays CDs capped with branched PEI. It has found applications in chemical sensing of metal ions, as it can selectively recognize Cu^{2+} ion and give a very sensitive signal response [202].

Another one-step synthesis of PEI-conjugated CDs by microwave-assisted pyrolysis method using glycerol (as CD precursor) and branched PEI25k resulted in the formation of CDs that was surface passivated with PEI. This CDs-PEI molecule was a water-soluble nitrogen-rich compound that passivated the surface to enhance the fluorescence and was also used as a polyelectrolyte to condense DNA. Another interesting part is that the DNA condensation capability and cytotoxicity of CDs-PEI can be regulated by pyrolysis time, which may be due to the destruction of some of the PEIs during the formation of CDs-PEI. This conjugate shows reasonably good gene expression of plasmid DNA in COS-7 cells and HepG2 cells as compared to control PEI25k [188]. The CDs-PEIs (internalized into cells) display tunable fluorescent emission under varying excitation wavelength, thus can have potential application in gene delivery and bioimaging.

PEI-conjugated CDs can also be synthesized by a two-step solvothermal method, i.e., first CDs are synthesized in water-glycol medium using glucose as CD precursor and attached PEI, which improves fluorescence quality by 300 fold, with a quantum yield of 3.5%. Han *et al.* [217] have conjugated it with CEA8 antibody and used it for labeling and imaging HeLa cells *in vitro*. It was not cytotoxic and has much better biosafety property than CdTe QD. These PEI-conjugated CDs could also be used for *in-vivo* applications.

When hyperbranched PEI is conjugated with CDs, it shows multistimuli-responsive property [39]. Further addition of isobutyric amide (IBAm) groups to CDs-PEI through the amidation reaction of isobutyric anhydride and the PEI moiety,

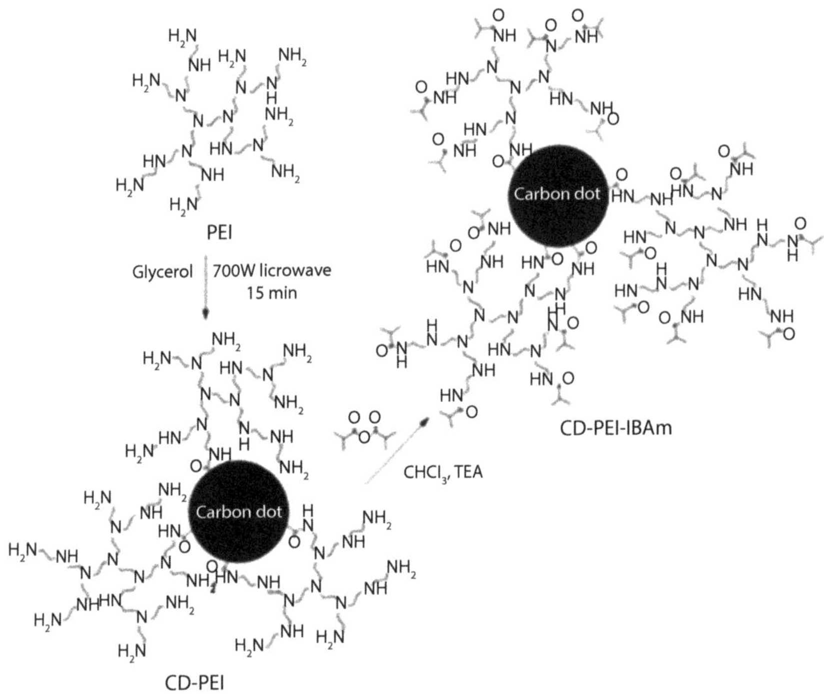

Figure 3.9 Multistimuli-responsive CDs prepared by Yin *et al.* [189].

produced CDs-PEI-IBAm's (Figure 3.9). This new conjugate is thermoresponsive as well as responsive to inorganic salt, pH and loaded organic moiety. Yin *et al.* [189] have modulated the cloud point temperature (Tcp) of the aqueous solutions of CDs-PEI-IBAm's in a broad range by changing the number of IBAm units of CDs-PEI-IBAm or varying the type and concentration of the inorganic salts, pH, and loaded organic molecules. However, the photoluminescence of CDs-PEI-IBAm's conjugates did get slightly or negligibly influenced by the added salts, pH, and the encapsulated organic guests.

Multi-functionalized CDs derived from glucose and PEI have shown both antibacterial (to both Gram-negative and Gram-positive bacteria) and gene delivery properties. For gene delivery, Dou *et al.* [218] investigated the expression of CDs condensed plasmid DNA in HEK 293T cells based on Luciferase expression assay. Gene transfection capability of these quaternized CDs was 10^4 times more efficient than the naked DNA delivery.

Multi-functionalized CDs, i.e., folate-conjugated reducible polyethylenimine passivated CDs(fc-rPEI-CDs), can simultaneously monitor the real-time response of therapy as well as that used for gene delivery for lung cancer therapy. Such conjugates emit visible blue photoluminescence under 360 nm excitation and can encapsulate multiple siRNAs (EGFR and cyclin B1) that can be released intracellularly in reductive environment; they (fc-rPEI-CDs) are highly biocompatible and a suitable siRNA gene delivery carrier for targeted lung cancer therapy, thus it (fc-rPEI-CDs/pooled siRNAs) can be selectively accumulated in lung cancer cells through receptor-mediated endocytosis, resulting in better gene silencing and anti-cancer effect [219].

3.3.3.3 CDs Conjugated with α-Cyclodextrin

As the name suggests, cyclodextrins are sugar molecules bound to a cyclic ring of oligosaccharides. α-Cyclodextrin is a polysaccharide of six glucose units that are covalently attached end to end via α-1,4 linkages. It is a soluble dietary fiber.

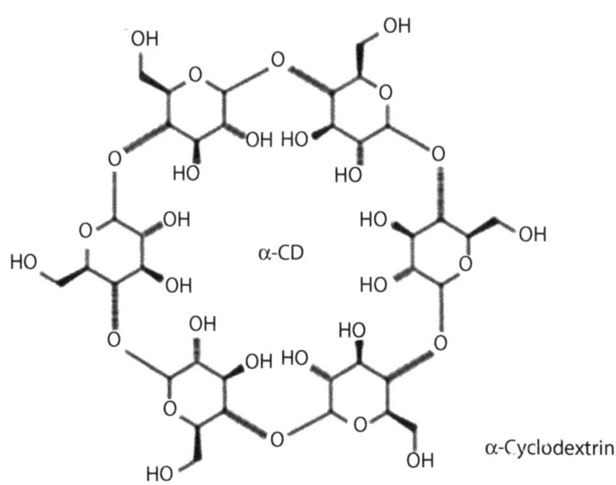

α-Cyclodextrin

α-Cyclodextrin in conjugation with CDs has been used in molecular recognition and optical sensing with the help of electron transfer mechanism. An approach to assemble α-cyclodextrin CDs having surface-encapsulated different derivatives of methyl viologen (MVe2+ as electron acceptors) has been tried to increase the photoinduced electron transfer (PET). The highest PET efficiency was observed when DHMV2+ derivative of methyl viologen was used. This aggregate has tubular formation. The advantage of this conjugate was that its time-resolved fluorescence decay and transient absorption spectroscopy study provided conclusive insight into the electron transfer processes at different time scales [220]. This study may find some applications in light energy conversion.

3.3.3.4 CDs Conjugated with Cysteamine

Cysteamine is an aminothiol containing an amine and a thiol functional group. It is synthesized in humans and other mammals by degradation of coenzyme A. It is the biosynthetic precursor to the neurotransmitter hypotaurine [221]. It is used to treat cystinosis.

Sharon's group [71] has prepared CDs from phenylalanine and conjugated with cysteamine-HCl that work as a linker to attach haloperidol (an antipsychotic drug). Cysteamine formed a stable bond with haloperidol for its controlled release kinetics, which according to statistical model under physiological conditions showed that controlled release was due to the porous nature of the CDs, as well as the impact of CysHCl as the linker. Moreover, CDs can get selective access to cells due to their large interactions with biological membranes [74], avoiding unwanted internalization.

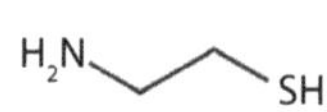

Cysteamine

3.3.3.5 CDs Conjugated with Dihydrolipoic Acid

Dihydroxy-lipoic acid is a powerful antioxidant. Fibrillation of protein is one of the disorders of neurodegenerative diseases such as Alzheimer's disease, Parkinson's disease and also type 2 diabetes. One of the main reasons for conjugation of dihydroxy-lipoic acid with CDs(DHLA- CDs) is to facilitate the water solubility of hydrophobic CDs. Leblanc's group [222] demonstrated the impact of CDs on human insulin protein fibrillation, as proteins may share a common mechanism to form fibrils. Their characterization efforts using thioflavin T fluorescence, circular dichroism spectroscopy and atomic force microscopy demonstrate that the presence of DHLA-CD increases the rate of human insulin fibrillation and accelerates the conformational changes of human insulin from α-helix to β-sheet. They suggested that this promoting effect may be due to the locally increased concentration of human insulin adsorbed on the surface of DHLA-CDs.

3.3.3.6 CDs Conjugated with Polyamidoamine (PAMAM) Dendrimers

Dendrimers are being extensively used as a multifunctional delivery vehicle. Dendrimers have well-defined size and three-dimensional architecture. Their high density of peripheral functional groups and a multivalent exterior makes dendrimers an attractive carrier for multiple conjugation reactions that can be simultaneously used for drug delivery and molecular imaging if they are conjugated with CDs [223–225]. Trials have been done using conjugates of CDs with anionic terminus and cationic acetylated G5 Poly(amidoamine) (G5-Ac85) dendrimers which were combined via noncovalent interactions to form self-assembled fluorescent hybrids. The presence of primary amine groups in the dendrimers enhances the fluorescence of the CDs, making them suitable for cellular imaging. It was used for encapsulating an anti-cancer chemo drug Epirubicin (EPI) in the interior of dendrimer that could be delivered intracellularly. Moreover, the *in-vitro* release of EPI from CDs@EPI⊂G5-Ac85 hybrids was faster in an acidic environment than under physiological conditions. This multifunctional CDs@EPI⊂G5-Ac85 hybrid serves as a dual-emission delivery system: i) it tracks the intracellular distribution using fluorescence microscopy and cellular uptake by flow cytometry and ii) cytotoxic effects of EPI drug is done by cell cycle analysis, field-emission scanning electron microscopy (FE-SEM), reactive oxygen species (ROS) generation and upregulation of apoptotic signaling genes. All these studies have shown induction of apoptosis by CDs@ EPI⊂G5-Ac85 hybrids in breast cancer (MCF-7) cells.

3.3.3.7 CDs Covalently Conjugated with Rhodamine B Dyes

The photoluminescent properties of CDs have a wide range of applicability in energy conversion, optoelectronics and sensing. Conjugation of CDs with organic dyes for resonance energy transfer process could enable the design of functional materials for enhancing the performance of solar cells, other optoelectronic devices, or to create new types of sensors. A CD with surface modified PEG_{1500N} was conjugated with Rhodamine B by covalent linkage to (RhB) isothiocyanate (CDs–PEG_{1500N}–Rh) (Figure 3.10). The spectroscopic analysis showed overlap between the emission spectra of CDs and the absorption spectrum of RhB molecules, which favored fluorescent (Förster)

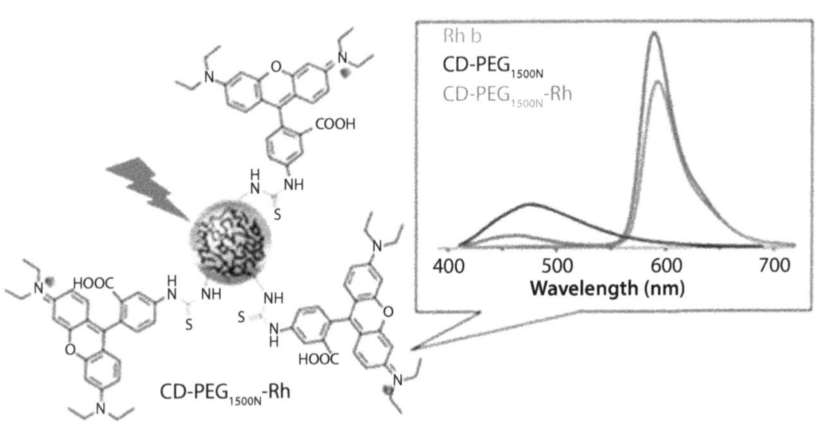

Figure 3.10 Covalent conjugation of Rhodamine B dyes with CD. (Adapted with permission from [140])

resonance energy transfer (FRET) from the CDs to the dye molecules. This conjugate may be used in FRET-based sensing platforms and photovoltaic devices [140].

3.3.3.8 CDs Conjugated with Fe⁻ Aminoclay (FeAC)

Aminoclay is composed of tetrahedral brucite (MgO) in the middle, sandwiched by octahedral silica (SiO_2) as the unit structure in the vertical direction (i.e., 2:1 trioctahedral clay) and a repeated tetrahedral/octahedral structure in pairs, knoSource:wn as the 1:1 dioctahedral structure. It is called *aminoclay* because it has diverse high-density primary amines [–$(CH2)_3NH_2$] in octahedral structures. CDs conjugated with nanoparticles of cationic Fe-aminoclay (FeAC) display cytotoxicity against HeLa cells but no cytotoxicity to normal cells up to concentration of 1000 μg/mL and can also be used for bioimaging because of the presence of CDs. FeAC NPs act as a substrate for CDs conjugation and help in the uptake of NPs into cancer cells due to the positively charged surface of FeAC NPs in physiological media. So, the CDs-FeAC NPs can be used as drug delivery carrier in cancer cells with simultaneous imaging [226].

3.3.3.9 CDs Conjugated with MWCNT

Carbon nanotubes (CNT) have emerged as an important matchless material in nanoscale science due to their unique properties such as smaller diameter (4 nm to >50 nm), long length (nanometer to centimeter), high mechanical strength, high thermal and chemical stability and good heat conduction. When a single layer of graphene sheet is rolled to form a cylindrical shape (Figure 3.11a), it is designated as single-walled carbon nanotubes (SWCNT). If more than one graphene sheet is rolled to form a hollow cylindrical shape, it is designated as multiwalled carbon nanotubes (MWCNT) (Figure 3.11b). In MWCNT, graphene sheets are separated by a distance of 3.4 Å, which is almost similar to the distance between graphene layers of graphite. They can be semiconducting or metallic depending

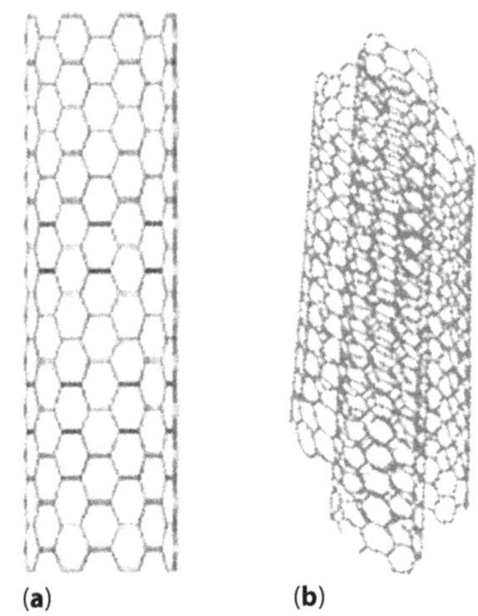

Figure 3.11 (a) SWCNT and (b) MWCNT.

upon the diameter of the tube. The number of walls, interaction between walls, disorderliness and the non-uniformity makes MWCNT a complex structure and provides MWCNTs with very good mechanical character with high Young's modulus of around one terapascal. Moreover, dangling bonds on the surface of MWCNT are good sites for functionalization.

Bai *et al.* [227] have decorated MWCNT with CDs for designing an enzyme-free electrochemical sensor for detection of hydrogen peroxide (H_2O_2) in living cells. When CD and MWCNT conjugate was modified by glassy carbon (GC) electrode, it performed better synergistic electrocatalytic activity towards hydrogen peroxide reduction as compared to CDs or MWCNT alone. This conjugate was successfully used for real-time tracking of H_2O_2 released from human cervical cancer cells. The enhanced electrochemical activity was due to the edge plane-like defective sites and lattice oxygen in the CD/MWCNT nanocomposites due to the small amount of decoration of carbon dots on the multiwalled carbon nanotubes (Figure 3.12).

3.3.4 CDs Conjugated with Antibiotics

Many antibiotics are on the market for killing pathogenic microbes. However, there are multiple problems that are faced

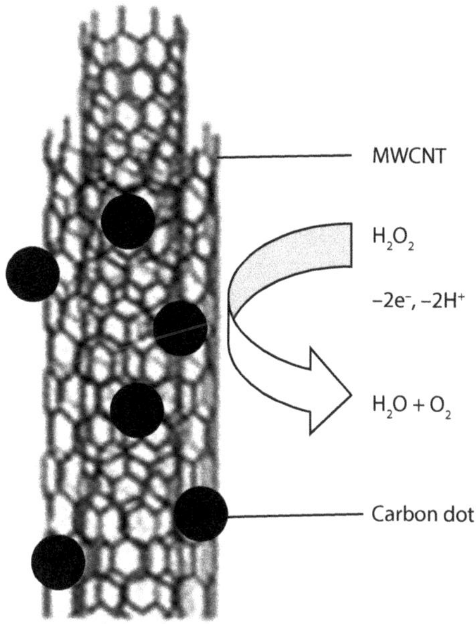

Figure 3.12 Schematic diagram of hydrogen peroxide detection by CD decorated MWCNT.

when an antibiotic is administered such as i) partial absorption of orally administered active ingredient, ii) quick renal flush out of part of them, iii) side effects due to heavy doses and iv) often development of resistant mutants to the drug, especially due to heavy doses. Some promising avenues to combat these problems have been shown by delivering antibiotics using functionalized CD conjugates as drug delivery vehicle. Quaternized CDs alone have selective antimicrobial activity against Gram-positive bacteria *Staphylococcus aureus* [228]. However, CD has been found to be a good molecule to conjugate drugs. Some of the examples of success of full loading and *in-vitro* and *in-vivo* delivery of antibiotics using CDs are discussed in brief.

3.3.4.1 CDs Conjugated with Ciprofloxacin

Ciprofloxacin is the most studied antibiotic for delivery using CDs. Ciprofloxacin hydrochloride is a broad-spectrum fluoroquinolone used as antibiotic. It restricts cell division by inhibiting DNA gyrase, type II topoisomerase and topoisomerase IV. It is used for treatment of both Gram-negative and Gram-positive infection of skin, urinary tract, respiratory tract, bone joints, abdomen, etc. Since it is a frequently used antibiotic often resulting in overdosing, causing resistance, an antibiotic conjugation strategy is an important need for controlled releases [229, 230].

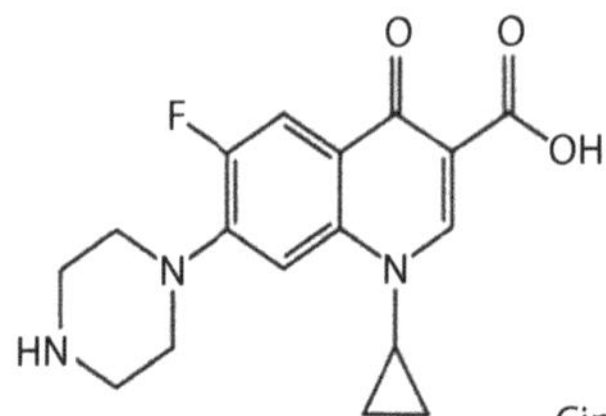

Ciprofloxacin

Gum arabic has been found to be a good precursor of CDs. An attempt to synthesize CDs using gum arabic, an edible carbon-rich material, by microwave-assisted method and to directly load ciprofloxacin has been tried by Sharon's group [72] (Figures 3.13 and 3.14). This biocompatible Cipro-CD conjugate i) is found to have >90% drug loading capacity, showing it to be an ideal vehicle for ferrying a significant amount of

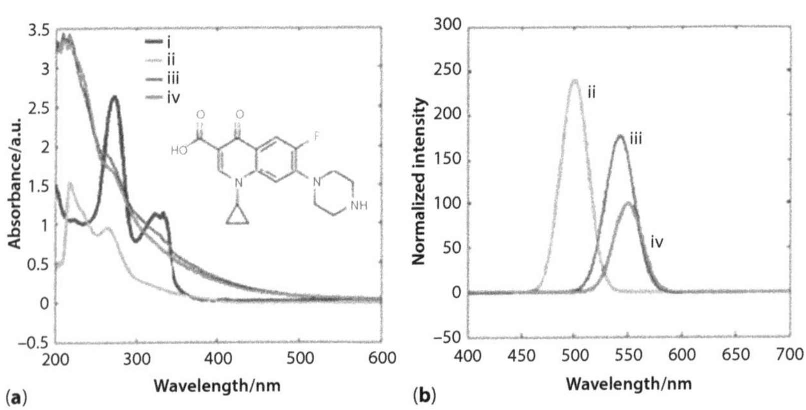

Figure 3.13 UV-Vis spectra of (i) Ciprofloxacin, (ii) CD, (iii) Cipro-CDs conjugate, (iv) Post dialysis Cipro-CD conjugate and (b) PL spectra of respective samples. (Adapted with permission from [72])

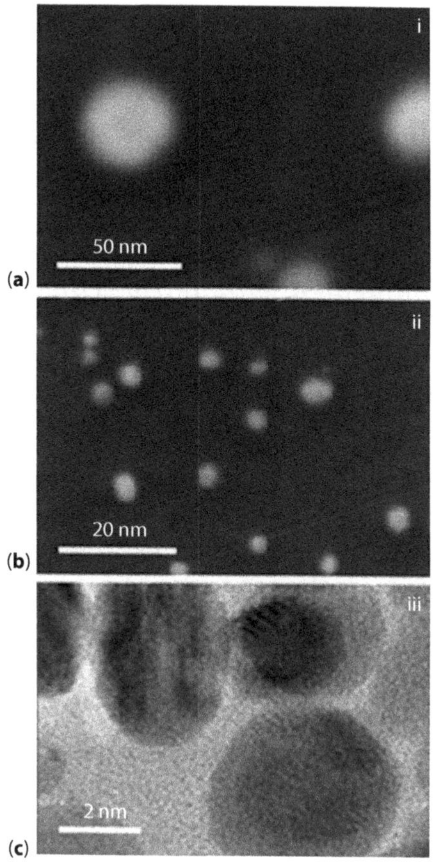

Figure 3.14 (b-i and b-ii) SEM images and (b-iii) HRTEM image of carbon dot synthesized using gum arabic. (Adapted with permission from [72])

clinical payloads; ii) releases ciprofloxacin under physiological conditions in a much regulated way over 24 h, making it act as ideal sinks to control pathogenic infections; iii) is biocompatible on Vero cells as compared to free ciprofloxacin (1.2 mM) even at very high concentrations; iv) shows enhanced antimicrobial activity against both model Gram-positive and Gram-negative microorganisms; v) is used for microbial imaging of the eukaryotic model—*Saccharomyces cerevisiae* (yeast); vi) is an efficient new nanocarrier for controlled drug release; vii) has high antimicrobial activity; and viii) results in ciprofloxacin being released at a sustained rate from the surface of CDs following the Higuchi model under physiological conditions.

Krishna *et al.* [172] have used PEG-diamine capped 2–5 nm diameter CD conjugated with glutamic acid and ciprofloxacin for simultaneous tracking and delivery of ciprofloxacin to control *E. coli* infection.

3.3.4.2 CDs Conjugated with Tetracycline

Tetracycline is a broad spectrum of antibiotic having bacteriostatic activity against pathogenic aerobic and anaerobic and both Gram-positive and Gram-negative bacteria. However, some microbes are intrinsically resistant to tetracycline such as *Pseudomonas aeruginosa* and *Proteus* spp.

CD (2.2 nm diam) conjugated with tetracycline alone and associated with β-cyclodextrin coated with calcium alginate beads, shows good stability under ambient conditions and at room atmosphere for > 60 days. Moreover, calcium alginate + CDs shows better swelling in comparison to only calcium alginate beads [231] and better loading capacity of the drug than the only calcium alginate. When tetracycline is associated with β-cyclodextrin it shows much greater loading efficiency (as high as 90%) for calcium alginate + CDs. The delivery of tetracycline is pH dependent. Maximum release is obtained at pH 1, emphasizing its suitability for the gastrointestinal tract where pH is low. Use of β-cyclodextrin in tetracycline + CD + calcium alginate renders slow and long-term drug administration.

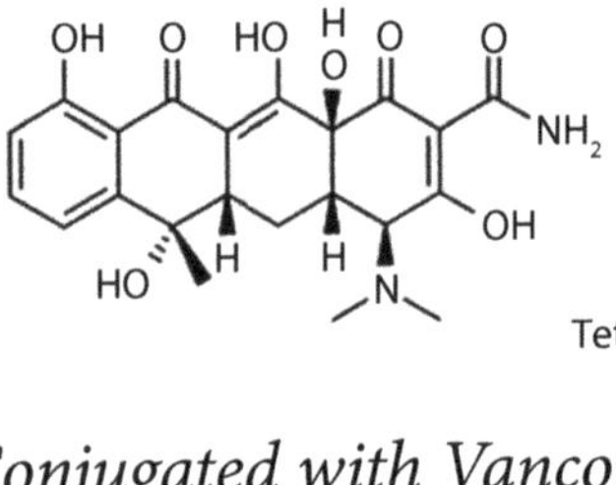

Tetracycline

3.3.4.3 CDs Conjugated with Vancomycin

Vancomycin is a tricyclic glycopeptide ($C_{66}H_{75}Cl_2N_9O_{24}$). Its mechanism of action in living cell is by blocking the construction of a cell wall, thus killing the cells. It is derived from soil actinomycetes known as *Amycolatopsis orientalis*, also known as *Streptomyces orientalis*. It is effective against Gram-positive bacteria, especially against staphylococci resistant to methicillin.

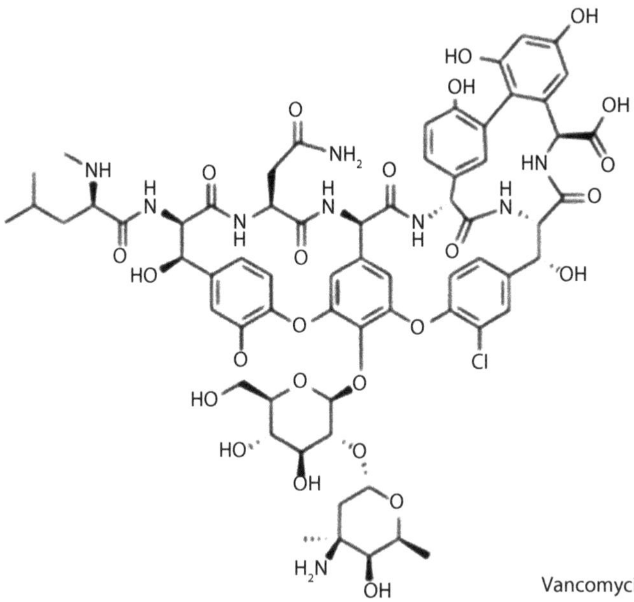

Vancomycin

Use of CDs for fast and accurate sensing of Gram-positive bacteria, including *S. aureus*, *B. subtilis* and *L. monocytogenes*, has been attempted by Zhong *et al.* [232]; they synthesized 2.2 nm diameter CDs from citric acid and urea and covalently linked the CDs to vancomycin.

3.3.4.4 CDs Conjugated with Ampicillin

Ampicillin is a broad-spectrum amino-penicillin group of antibiotic. It is a β-lactam antibiotic, used against many Gram-positive and Gram-negative bacteria. Ampicillin was the first "broad spectrum" penicillin. It can penetrate Gram-positive and some Gram-negative bacteria as it has an amino group, which helps the drug penetrate the outer membrane of Gram-negative bacteria. Ampicillin is bacteriolytic because it acts as an irreversible inhibitor of *transpeptidase* (an enzyme needed by bacteria to make the cell wall). It inhibits the third and final stage of bacterial cell wall synthesis in binary fission, thus leading to cell lysis.

Recently, Jijie *et al.* [233] have reported high effectivity of CDs-conjugated ampicillin in inactivating the growth of *E. coli* in the presence of visible light where moderate quantities of reactive oxygen species are generated, which results in additional photodynamic inactivation of the microbe.

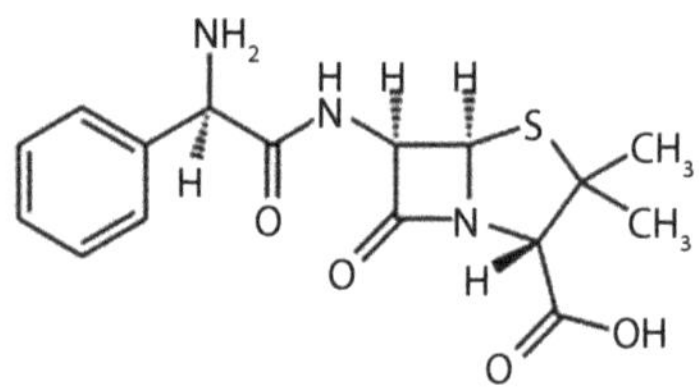

Ampicillin

3.3.4.5 CDs Conjugated with Streptomycin

The antibiotic streptomycin is derived from *Streptomyces griseus* and is used against many pathogenic bacteria.

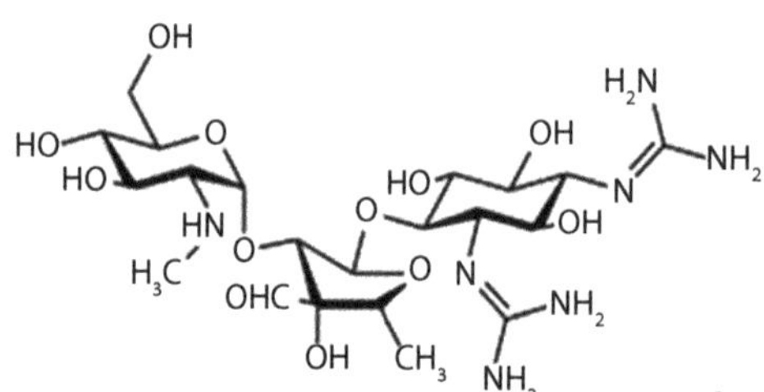

Streptomycin

Streptomycin is an amino-glycoside antibiotic. It is used to inhibit growth of both Gram-positive and Gram-negative bacteria by binding irreversibly to bacterial ribosome (small 16S rRNA of 30S subunit of ribosome), thus inhibiting protein synthesis. It can also kill algae and fungi. Streptomycin is administered in both plants and animals to control microbial infection. Streptomycin has been used as precursor to synthesize nitrogen-doped CDs, but not enough work has been initiated on the conjugation of CDs with streptomycin. Efforts are underway in our lab to conjugate streptomycin with CDs derived from gum arabic for foliar application to deliver streptomycin for trans-phloem bacterial disease, i.e., apple blight caused by *Erwinia* spp. The drug loading efficiency was found to be 75.24%. To study the drug release kinetics, *in-vitro* drug release was carried out under physiological condition in PBS (pH 5.7) at 30 °C using the dialysis method. Released drug was assessed by absorbance. *In-vitro* drug release data is given in Table 3.1.

To predict the kinetics and mechanism of drug release, the results of the *in-vitro* drug release study were fitted with various kinetic models like Zero order, First order, Higuchi model and Hixson-Crowell's cube root model. The criteria for selecting the most appropriate model were based on high correlation coefficient (r^2) value in various models (Table 3.2). The Hixson-Crowell model that gave higher 'r^2' value was considered as best fit model.

Table 3.1 % Streptomycin released *in vitro* at different time intervals.

Time/Hour	% Drug released
1	3.22
2	8.44
3	20.26
24	27.63
48	31.67
78	48.21

Table 3.2 Summary of values of coefficient correlation (r^2) at pH 5.7 as observed in all the models.

Models	r^2 Values
Zero model	0.920
First order	0.966
Higuchi	0.872
Hixson-Crowell	0.973

The Hixson-Crowell release equation is

$$3\sqrt{Q_0} - 3\sqrt{Q_t} = K_{HC}.t \quad (3.1)$$

Where:

Q_0 = Initial amount of drug
Q_t = Cumulative amount of drug release at time "t"
K_{HC} = Hixson-Crowell release constant
t = Time in hours.

It describes the drug releases by dissolution and with the changes in surface area and diameter of the particles or tablets.

Chemical bonds between CDs and streptomycin conjugate as proposed by Sharon are shown in Figure 3.15.

3.3.5 CDs Conjugated with Anti-Neurodegenerative Drugs for Delivery to Central Nervous System

For delivery of drugs to the central nervous system, it is necessary to pass the blood-brain barrier (BBB) that is organized

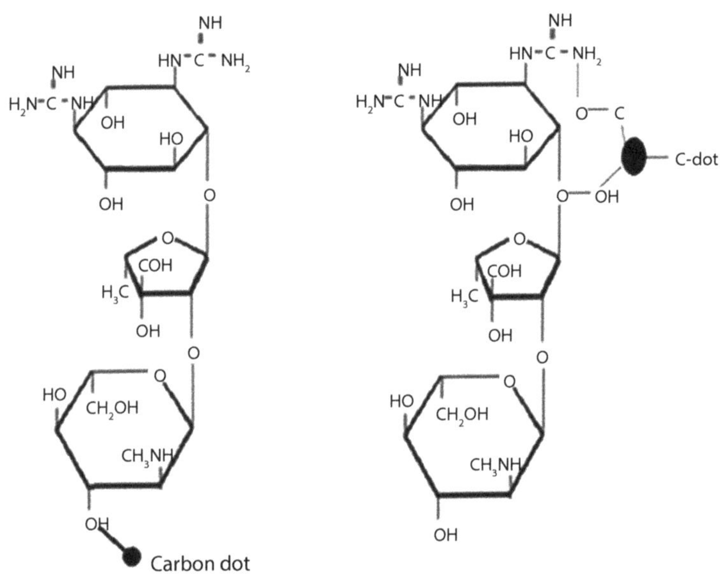

Figure 3.15 Chemical bonds between CDs and streptomycin conjugate proposed by Sharon.

with the capillary endothelial cells. These cells hinder the penetration of both big (such as therapeutic agents like antibodies, peptides, proteins, etc.) and also small molecules of drugs [234, 235]. This obstruction due to BBB has resulted in the use of increased concentration of drugs and overdosing to achieve the desired effect. No wonder there are excessive side effects faced by patients suffering from neurodegenerative diseases. To overcome this, particles of nanometer size have been considered because the permeability coefficient (Papp) of nanoparticles of <20 nm is higher; also, amino-Q-dots were found to be more actively transported through the *in-vitro* BBB model than other surface charged Q-dots.

As discussed earlier, CDs are novel and excellent drug delivery nanocarriers because their particle size and surface properties can be manipulated to circumvent rapid clearance by phagocytic cells, allowing both passive and active drug targeting. Moreover, CDs have the ability to reach the smallest capillary vessels, due to their tiny volume, and to penetrate the tissues either through the para-cellular or the trans-cellular pathways.

Common neurodegenerative disorders caused by schizophrenia, Alzheimer's disease, brain tumors, etc., need drug delivery to the brain. However, there are disorders which are due to improper functioning of brain cells, such as imbalanced production of transferrin, which also demand therapeutic agents to be delivered to the brain. CD conjugates to treat neurogenerative disorders can be administered through oral, pulmonary, nasal, parental, and intraocular routes.

3.3.5.1 CDs Conjugated with Haloperidol

Haloperidol [4-(4-(p-Chlorophenyl)-4-Hydroxypiperidino)-4'-fluorobutyrophenone] is a butyrophenone derivative with antipsychotic properties used for alleviating the symptoms of schizophrenia and manic states. Haloperidol is a tertiary amine that tends to form interphase between water/air or water/lipid at very low concentrations of the order of 10–7 M. It is a neuroleptic drug. The mechanism of action of haloperidol is attributed to the inhibition of the transport mechanism of cerebral monoamines, particularly by blocking the impulse transmission in dopaminergic neurons.

Sharon's group [71] has focused on the delivery of a model neuroleptic compound, haloperidol, via carbon dots made from phenylalanine using microwave-assisted method and studying the *in-vitro* drug release profile and kinetics under physiological conditions. To achieve the intricate production of CDs bound haloperidol, the drug was covalently linked via amide linkage to CDs through cysteamine hydrochloride (CysHCl). Microwave synthesized CDs have great potential as drug delivery vehicles because of their small size, ease of synthesis, and their hydrophilic outer shell.

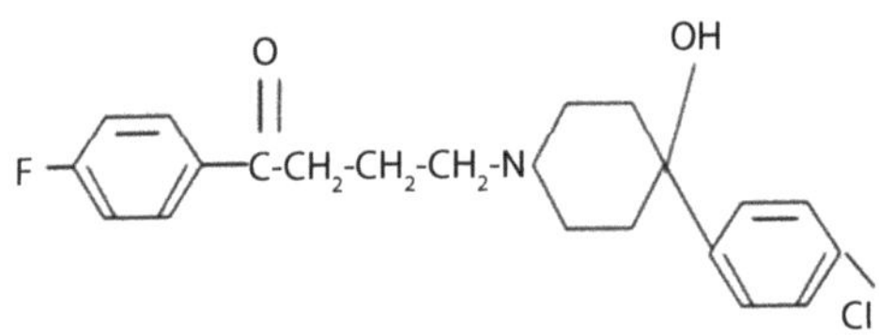

Haloperidol ($C_{21}H_{23}CLFNO_2$)

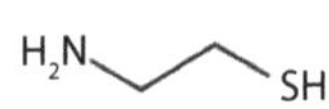

Cysteamine

Cysteamine is a decarboxylated product of cysteine and is the most stable amino-thiol. It is used as a linker because it has many therapeutic applications in the treatment of cystinosis, radiation sickness, and also as a potent depletor of prolactin concentration in the pituitary. It has been used in studies on oxidative stress, glycation, as a substrate to study the specificity and kinetics of cysteamine (2-aminoethanethiol) dioxygenase. Cysteamine forms an important biochemical coenzyme A by combining with pantothenate and adenosine triphosphate. Research on CysHCl suggests its role in increasing the somatotropin secretion in several species, possibly through inhibition of somatostatin (produced in the hypothalamus, it inhibits the pituitary gland's secretion of growth hormone and thyroid-stimulating hormone) secretion. In drug delivery system, CysHCl is used as ligand for surface functionalization of different nanomaterials. The amine and thiol portion of the molecule are involved in its ligand acting property.

The phenylalanine synthesized CDs were functionalized with cysteamine hydrochloride (CysHCl), which acts as linker for the attachment of haloperidol to form CD-CysHCl complex for efficient delivery across the BBB. Attachment of haloperidol with CD-CysHCl conjugate was confirmed by UV spectrophotometry (Figure 3.16) and FTIR (Figure 3.17) analysis. Most

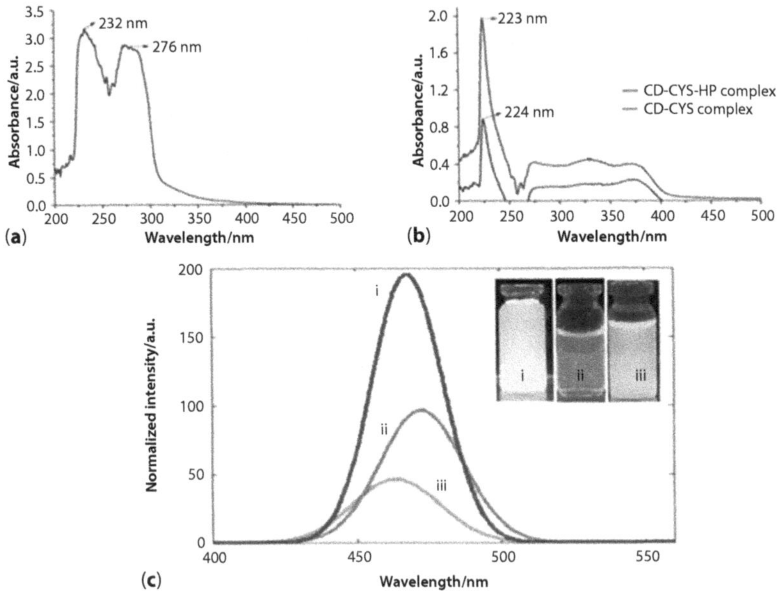

Figure 3.16 UV-Visible spectra of (a) CDs synthesized from phenylalanine, showing sharp peak at 232 nm and a broad peak at 276 nm, confirming the formation of CDs and (b) CD-CysHCl and CD-CysHCl-HP complex, with respect to peaks observed in the case of CD-CysHCl complex, there was a slight blue shift of 1 nm (from 224 to 223 nm) and altered humps at 273, 330 and 375 nm specific to HP and CD-CysHCl-HP complex. The peak intensity is elevated, which is due to masking of functional groups and attachment of HP onto the surface of CDs. (c) Photoluminescence spectra of CDs, CDs-CysHCl and CDsCysHCl haloperidol complex; labeled as i, ii and iii respectively. Inset showing the same under UV light. (Adapted with permission from [71])

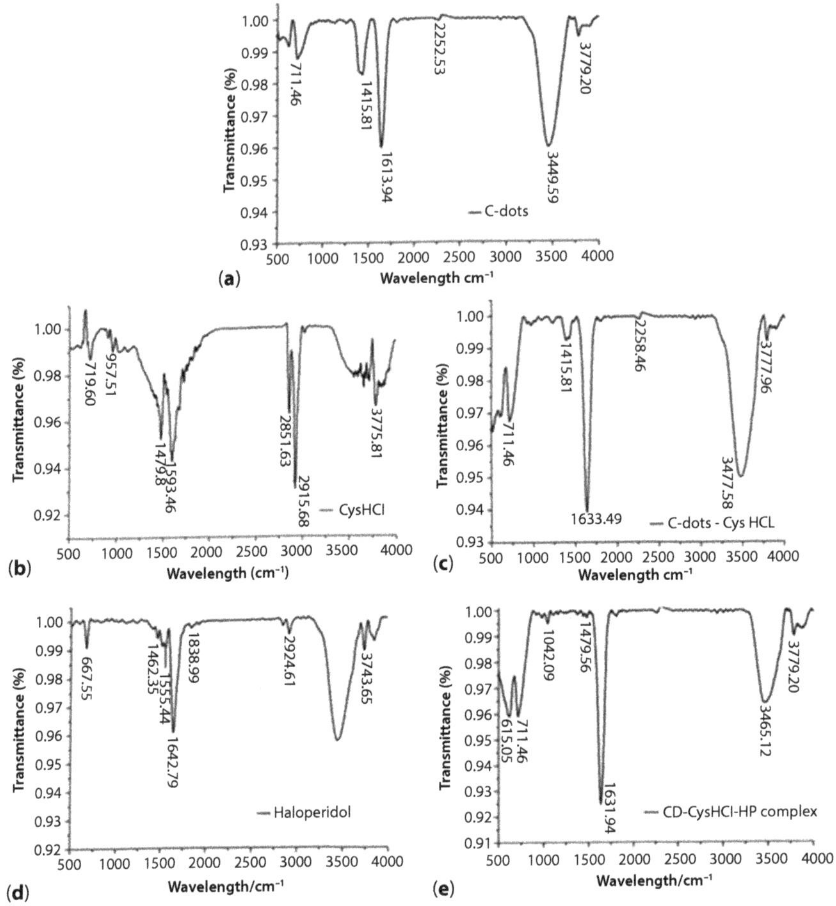

Figure 3.17 FTIR spectra of (a) CDs, (b) Cysteamine-HCl, (c) Haloperidol, (d) Conjugate of CD and Cysteamine-HCl, (e) Conjugate of CD+ Cysteamine-HCl + Haloperidol. (Adapted with permission from [71])

of the CDs, prepared by microwave-assisted heating, or supported method, have an absorption band around 260–320 nm. The appearance of sharp peak (carbon dots synthesized from phenylalanine) at 232 nm is due to $\pi \rightarrow \pi^*$ electronic transition, which is associated with C=C groups on the surface of CDs. Another hump observed at 276 nm is due to $n \rightarrow \pi^*$ electronic transition occurring due to the presence of carbonyl groups. As no background absorption is seen in the visible range, it assures the lack of other conjugate moieties and carbonaceous material which shows characteristic absorption within the visible range.

FTIR of CDs (Figure 3.17a) – a sharp peak at 1613 cm^{-1} (C=O vibrational stretch) and a broad peak centered at 3449 cm^{-1} (O–H vibrational stretch) reveal the existence of carboxylate and hydroxyl groups on the surfaces of the CDs. A C–O vibrational stretch was detected, which indicates that the surfaces of the CDs are partially oxidized. In addition, the peak observed at 711 cm^{-1} can be assigned to the methylene skeletal C-C vibrations and presence of $-CH_3$ groups.

The FTIR spectra of CDs, Cysteamine-HCl, Haloperidol, C–dot-CysHCl conjugate and C–dot-CysHCl-Haloperidol conjugate are shown in Figure 3.17a–e.

FTIR Spectra of CD functionalized with CysHCl – exhibits a shift and increased intensity of peak at 1633 cm^{-1} (-NH vibrational bending) and decline in the absorption of peak located at 1415 cm^{-1}; as compared to only CD. (C-C stretch, O-H bend) confirms the grafting of CysHCl moieties onto the surface of CDs. The results suggest that CysHCl must have anchored to CDs via an amide linkage.

FTIR Spectra of CD-CysHCl-HP conjugate – As seen in Figure 3.15c (FTIR of Haloperidol), the absorption peak at 1462 cm^{-1} and 1555 cm^{-1} corresponds to C=C vibrational stretch and presence of tertiary amine group. The sharp peak observed at 1642 cm^{-1} corresponds to ketone group. In addition, the peaks observed at 667 cm^{-1} and 1838 cm^{-1} correspond to the presence of chlorine and fluorine groups respectively.

A comparison of Figure 3.17c, d and e shows the appearance of a new peak at 1042 cm^{-1}, which indicates a C-O/C-N/S=O stretch. The peak observed at 615 cm^{-1} corresponds to C-CL stretch specific to haloperidol. In addition, the peak observed at 3465 cm^{-1} corresponds to N-H/O-H stretch. The diminishing of peak around 1400–1500 cm^{-1} indicates the grafting of HP moieties onto the surface of CDs. The above results suggest that haloperidol must have anchored onto the surface of CDs via amide linkage or hydrogen bonding.

This complex has shown that a drug loading efficiency of 75.24% and *in-vitro* drug release efficiency at pH 7.2 in 72 h was >50%.

3.3.5.2 CDs Conjugated with Transferrin

Transferrin (mol wt ~80 KDa) is a glycoprotein. It reversibly binds to iron by two specific high-affinity Fe(III) binding sites. Human transferrin is composed of a polypeptide chain having 679 amino acids and two carbohydrate chains. There are two domains in this protein α-helices and β-sheets that form two domains. Human transferrin is encoded by the *TF gene.* Apart from the liver, the brain also produces transferrin. Transferrin absorbs iron from duodenum and WBC and then distributes it to all tissues. Transferrin imbalance in the body is the cause of iron deficiency or iron overload disorders, such as hemochromatosis, because the main physiological role of transferrin is in iron transportation in the circulatory system [236]. However, one of the properties of transferrin is that it is a promising cancer-targeting agent in multiple delivery systems because of the overexpression of transferrin receptors at the surface of many cancer cells Quian *et al.* [237]. Therefore, transferrin is being used as a molecular Trojan horse for targeting selected cancer cells [238–241].

Li *et al.* [242] have shown that CDs conjugated with transferrin can target cancer cell, and they implied that due to the presence of biocompatible CDs, this conjugate can be used for *in-vitro* cancer diagnosis.

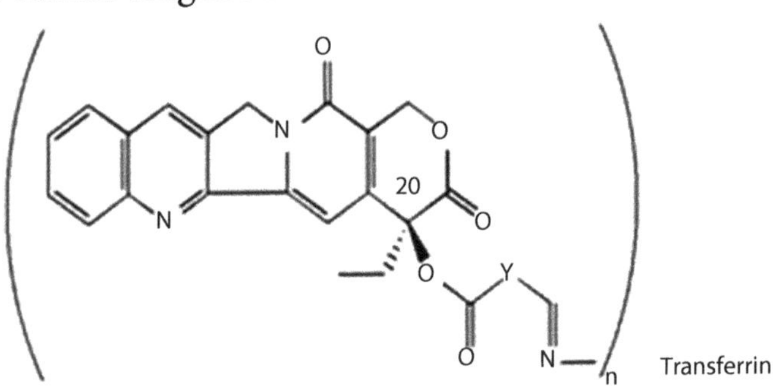

Li *et al.* [242] have utilized the CDs-conjugated transferrin receptor mediated endocytosis through BBB receptors for increased efficacy and selective targeting of doxorubicin to pediatric brain tumors. They used CD–transferrin–doxorubicin covalent conjugate (CDs–Trans–Dox) for *in-vitro* studies. The *in-vitro* results indicated that there was higher uptake of the CDs–Trans–Dox conjugate as compared to Dox alone. The conjugate reduced the viability of pediatric brain tumor cell lines by 14–45% more in all pediatric brain tumor cell lines. The brain tumor cells have increased requirements for iron; therefore, they express a high level of transferrin receptors, resulting in uptake of more transferrin bound chemotherapeutic agent than the noncancerous cells, allowing for a more targeted delivery system.

In another experimental trial, Li *et al.* [243] used a zebrafish model to study CDs and transferrin conjugate for crossing the BBB and entering the central nervous system by injecting them into the vasculature of zebrafish. CDs have been known to inhibit protein fibrillation and deliver drugs. Therefore, CDs and CDs-conjugated transferrin were tried for treatment of central nervous system (CNS) diseases. It was observed that the transferrin-CDs conjugate could enter the CNS while CDs alone did not.

Zhang *et al.* [108], in an attempt to find out the binding between CD and transferrin, have synthesized glycine-modified 10 nm CDs with high QY (31%) by microwave-assisted method using citric acid and glycine. They found that glycine-modified CD has strong interaction with human transferrin (hTf) and there is one type of binding site between them.

3.3.5.3 CDs Conjugated with Curcumin

Major pathological factors in the initiation and promotion of neurodegeneration in Alzheimer's disease are: i) Amyloid plaque formation between nerve cells in the brain, along with degenerating bits of neurons and other cells causing oxidative damage and impairment of the synapse connections. Amyloid plaque causes loss of synaptophysin, a protein involved in brain synapses. ii) Tau protein, which functions as microtubule, is transformed abnormally, causing twisted filaments in neurons, resulting in the collapse of the neurons' transport system, impairing communication between nerve cells. iii) Gliosis, i.e., abnormal growth and proliferation of neuroglial cells, which provide mechanical support and electrical insulation between neurons that gets impaired. iv) Abnormal proliferation of astrocytes.

Curcumin is a diarylheptanoid curcuminoid that is a natural phenol that occurs in *Curcuma longa*. Curcumin's ability to scavenge free radicals and suppress inflammatory cytokines seems promising to prevent and treat Alzheimer's disease disorder. Curcumin is known to a) lower proinflammatory cytokine known as interleukin-1 beta, b) act as inflammatory biological marker, c) glial fibrillary acidic protein, d) suppress loss of synaptophysis, e) directly binds to beta amyloid to block aggregation and fibril formation, f) suppress neuroglial proliferation, g) inhibits astrocytosis, h) acts as an antioxidant, which induces a heat-shock response in the brain as an advantageous effect against oxidative damage.

However, the blood-brain barrier and neurotoxicity of many traditional metal nanoparticles as drug delivery vehicle have limited their utility in Alzheimer's disease or other neurodegenerative disorders.

To circumvent this, conjugation of curcumin with CDs, as they can cross the BBB, appears to be a good solution because we could get curcumin attached to NaOH passivated CDs. The NMR analysis reveals that there is strong interaction of negatively charged CDs with positively charged cationic groups of curcumin. FTIR measurements further confirmed the surface modification by curcumin. So far, the effects of curcumin attached CDs against damage induced by beta amyloid has been studied only *in vitro*. Hence, there is a possibility of using

CD + Curcumin conjugate to deliver curcumin to the central nervous system, but it cannot be used for diagnostic purpose to trace the transport of CD conjugate in the brain, because, as found by Shi *et al.* [244] and Zhang *et al.* [245], attachment of curcumin to CDs quenches its photoluminescence.

Shi *et al.* [244] have shown that there is a fluorescence resonance energy transfer (FRET) between CDs and curcumin (Cur); based on this fact they designed a sensing system for detecting curcumin (Figure 3.18). Due to FRET, the CD (having strong blue fluorescence and excitation wavelength-dependent emission behavior with the maximum excitation and emission wavelength at 360 nm and 420 nm, respectively) prepared by them using DTPA (diethylenetriaminepentaacetic acid) as the carbon source, was quenched with the introduction of curcumin. The decrease in intensity was linearly proportional to the concentration of Cur in the range of 0.74–5.18 μg mL^{-1}. Hence, they implied that quantitative detection of curcumin with a detection limit of 44.8 ng mL^{-1} is possible.

Similar quenching was observed by Zhang *et al.* [245] when they used nitrogen-doped CDs synthesized using citric acid as precursor and urea as source of nitrogen by hydrothermal method. They suggested the quenching of CDs due to inner filter effect. They applied these nitrogen-doped CDs to detect curcumin in urine and could recover 95.71–103.81% curcumin.

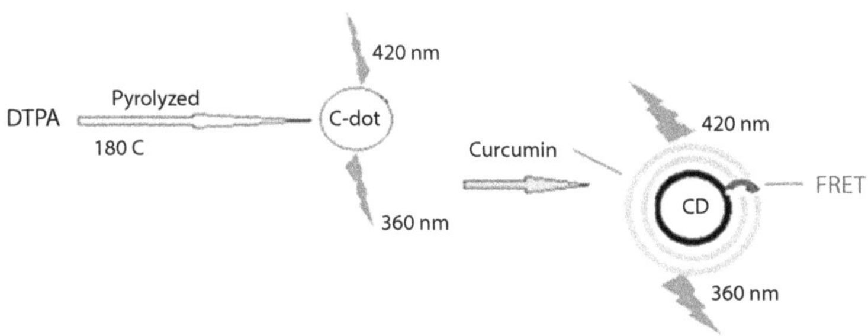

Figure 3.18 Schematic diagram showing quenching of CDs fluorescence due to FRET between CDs and curcumin. (Adapted with permission from [244])

3.3.6 CDs Conjugated with Anticancer Drugs

Use of CD conjugates for cancer therapy and diagnosis needs multiple functionalization of its surface in order to attach a functional group so that a particular anti-cancer drug can be attached that can be stored and delivered for treatment and diagnosis, keeping the fluorescence property of CDs undisturbed. Moreover, the conjugate has to be biodegradable so that it can be flushed out. CDs have been found suitable for this purpose, as their physical and chemical properties can be altered by conjugation. Many anti-cancer drugs, especially for malignant tumors, need nucleus-targeting drug delivery, as they enter the cell nucleus and damage the genes to stop proliferation of the cancer cells. Since CDs contain a sp^2- and sp^3-hybridized carbon, they can load aromatic drugs via strong π–π interactions. To allow CDs to enter the cell nucleus without being trapped in the cell cytoplasm, the size and surface charge plays an important role. The positively charged nanoparticles are preferentially internalized by cells, whereas the negatively charged ones interact with nuclei whose pH is consistently 0.3 to 0.5 units above that of the cytosol [246, 247].

Cellular uptake kinetics, uptake efficiencies, and intracellular distribution have been studied in three cancerous cell lines (breast – MDA-MB-231, bladder – AY-27, and brain – U87-MG) using flow cytometry and confocal microscopy. It was found that the intracellular distribution was almost the same in all three cell lines, but the uptake kinetics and efficiencies varied according to the cell morphology [248]. Apart from doxorubicin, cis-platin iii and carboplatin were also evaluated. Doxorubicin and cisplatin iii were found to be more toxic to cancer cells than carboplatin.

3.3.6.1 CDs Conjugated with Doxorubicin

Doxorubicin is a nucleus targeting anti-cancer drug used for chemotherapy by injecting through the vein. It is an anthracycline.

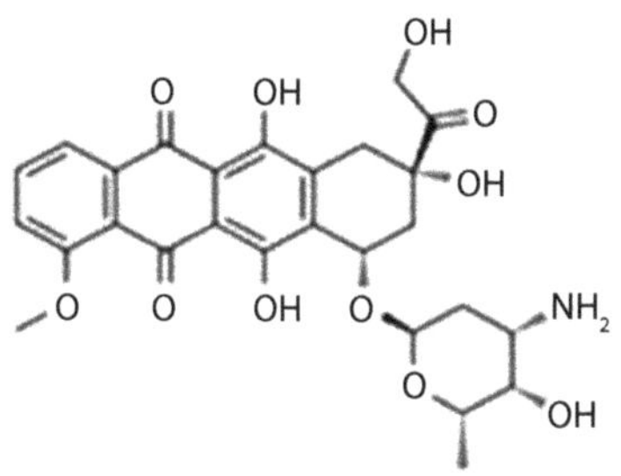

Doxorubicin

Different CD conjugates supporting the doxorubicin complex have been discussed earlier in this chapter, e.g., CDs conjugated with DNA and doxorubicin as a hybrid hydrogel for sustained release of doxorubicin [154]. CDs conjugated with GNR and doxorubicin act as a tri-pronged molecular vehicle for targeted (folic acid was used as synaphic targeting molecule) drug delivery, photothermal therapy and biological imaging, etc. [131].

Sorbitol-derived CDs have been explored for the delivery of therapeutic payloads for anti-cancer drug doxorubicin using folic acid (FA) as navigational molecule due to its high expression in most of the cancer cells. For attaching doxorubicin, the surfaces of CDs were protected with bovine serum albumin (BSA) to make them more biocompatible and hold high amount of drugs. The release profile of doxorubicin has shown that it follows first order standard statistical models at pH 7.2; and cellular imaging performed using epifluorescence microscopy showed bright green colored fluorescence due to internalization of CDs specifically targeted with FA in HeLa cells [50].

The preparation of CDs by conjugating with β-alanine acted as both passivating agent and zwitterionic ligand used for the delivery of doxorubicin to the nucleus. Doxorubicin was grafted onto the CD conjugate by noncovalent bonding. Moreover, multicolor fluorescence helped in optical monitoring of translocation of doxorubicin [249].

In an attempt to deliver doxorubicin to the nuclei of cancer cells, CD was functionalized with a nuclear localization signal peptide (NLS-CDs). The doxorubicin was conjugated with

NLS-CDs (DOX-CDs) through an acid-labile hydrazone bond. This bond is easily cleavable in the weakly acidic intracellular condition. Both cytotoxicity (of the human lung adenocarcinoma A549 cells) of the conjugate and the cellular uptake (by cell imaging) was monitored and confirmed using flow cytometry and confocal laser scanning microscopy. When the *in-vivo* therapeutic efficacy of DOX-CDs was tested in A549 xenograft nude mice model, the conjugate exhibited an enhanced ability to inhibit tumor growth as compared with free doxorubicin [250].

pH-Sensitive N-doped CD has been used for conjugation of two drugs at a time; one delivered to cytoplasm—heparin—and the other to nucleus—doxorubicin [251]. To conjugate both the drugs on the CDs, first heparin was attached via chemical bonding and then doxorubicin was loaded onto the CD-heparin conjugate by taking advantage of the electrostatic interactions between doxorubicin. Since the surface of CDs–Hep has a number of negative charges, it enabled the attachment of doxorubicin without any further modification. High drug loading was attained because of the surface amino group on the CDs. CD-heparin makes a stable conjugate, which was internalized, and subsequently doxorubicin entered the nucleus (Figure 3.19). The doxorubicin-heparin-CD conjugate shows accelerated rate of doxorubicin release in acidic buffer.

3.3.6.2 CDs Conjugated with Cisplatin

Cisplatin $[Pt(NH_3)_2Cl_2]$ is an intravenously administered anticancer drug. It interferes with DNA replication and kills the fast-growing cancer cells.

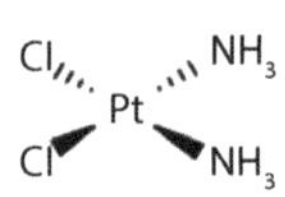

Cisplatin

Comparatively less work has been done on conjugating Cisplatin with CDs. Yang *et al.* [252] conjugated Pt^{IV}–azide-FA@CD for photocontrolled targeted platinum drug delivery. Folate

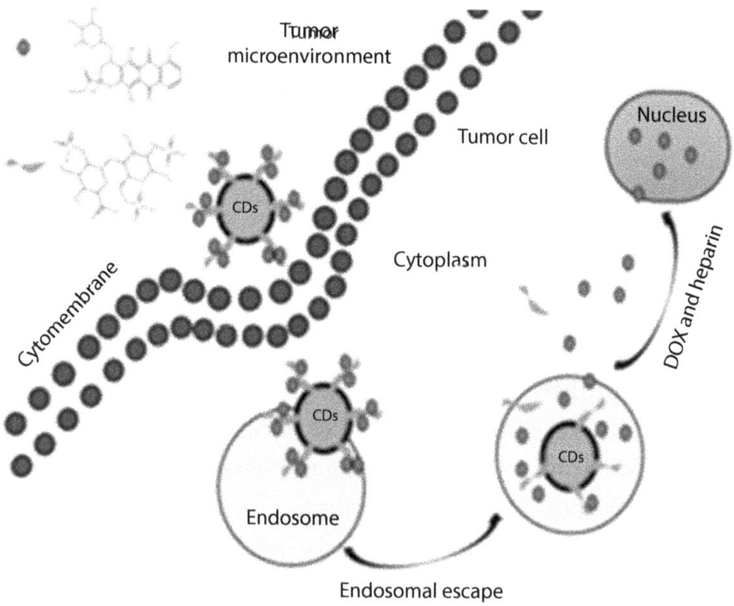

Figure 3.19 Schematic diagram of the intracellular release behavior of doxorubicin from the CDs–Hep/DOX drug delivery system. (Adapted with permission from [251])

(FA) was added as a targeting molecule to deliver drug to cancer cells. Moreover, a similar Cis-platin conjugate was also done. Simultaneously it was traced based on its inherent fluorescence. Both Pt^{IV}–azide-FA@CDs and Cis-platin-FA@CDs conjugates showed no dark cytotoxicity even in the presence of reducing glutathione, whereas both could get photoactivated by either UV or visible light, producing a similar level of cytotoxicity.

Oxaliplatin is another platinum-containing anti-cancer drug, used for treating colorectal cancer, that has been integrated with CDs to be used as a theranostic agent. CD-Oxaliplatin is synthesized by condensation reaction between the amino groups on the surface of CDs and the carboxyl group of the oxaliplatin derivative Oxa(IV)-COOH. This conjugate integrates the optical properties of CDs and the anticancer function of oxaliplatin, thus can be used for simultaneous drug delivery and fluorescent tracking [253].

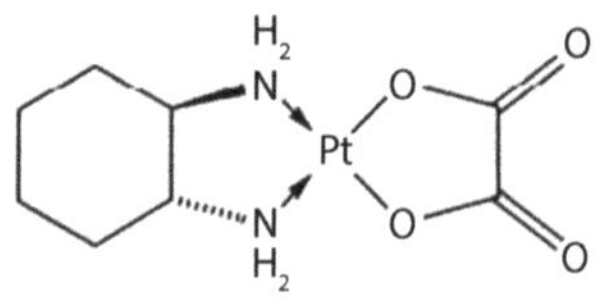

Oxaliplatin ($C_8H_{14}N_2O_4Pt$)

3.4 Summary

Carbon dots (CDs) are a new type of QD and have attracted a great deal of attention in recent years because of their promising applications in diagnostics, optoelectronic devices, pollutant detection and catalysis. In comparison with conventional inorganic QD, CDs have many advantages, including low toxicity, chemical inertness, biocompatibility and no photobleaching. In addition, typical CDs contain a large amount of hydrophilic carboxyl and hydroxyl groups, imparting them with excellent water solubility and multi-functions. For enhancing the efficacies of CD for various applications, CDs have been conjugated with inorganic, organic, biogenic and synthesized drugs and antibiotics. Though there has been a large amount of research work done on drug delivery using various C-dot conjugates, most of it is still at the bench scale and has not been commercialized.

4

CD as Drug Delivery Vehicle

"Research is the immersion into the unknown....
The 'stupidity' is an existential fact, inherent in our efforts to push our way into unfamiliar knowledge's. The more comfortable we become with being 'stupid,' the deeper we will wade into the unknown and the more likely we are to make big discoveries."
Martin A. Schwartz, 2008

4.1 Introduction

Conventional dug delivery has evolved and refers to approaches, formulations, technologies, and systems of delivering drug or therapeutic compound in the part of the body suffering from diseased cells. One of the concerns in drug delivery has been targeted delivery of a required quantity of drug and duration of drug presence. For delivery of drug, a particular chemical formulation suitable for the drug is used. However, drug delivery is also involved in medical devices or drug-device combination products and integration of dosage with route of administration. Other concerns in drug delivery are drug release profile, absorption, distribution and elimination for the benefit of improving product efficacy and safety, as well as patient convenience and compliance along with the delivery route, dosage, targeting and sustainable release profile.

The non-invasive pre-oral drug delivery routes are oral (through the mouth), topical (skin), transmucosal (nasal, buccal/sublingual, vaginal, ocular and rectal) and inhalation. Many therapeutic materials, such as peptides and proteins, antibodies, vaccines and gene-based drugs, are not delivered using these routes. This is because there is a chance of their enzymatic degradation in the route and failure in their efficient absorption into the systemic circulation due to molecular size and charge issues, in order for them to be therapeutically effective. Hence, protein and peptide drugs are injected. Targeted drug delivery, especially for cancer or neurodegenerative diseases, and sustained release formulations are also a part of the drug delivery system of present day treatment. Present day concerns have evolved a need to:

- Minimize the dosage so as to reduce the side effects.
- Release the drug over a period of time in a controlled manner.
- Avoid unnecessary use of formulation.
- Provide an efficient targeted delivery system to avoid the host's defense mechanisms and circulate drug to its intended site of action.
- Provide drug carriers or conjugates that are biodegradable to nontoxic material.
- Ensure that drug does not degrade prior to the release of the drug.
- Easily dispose of drug carrier after delivering the drug.

In conventional drug delivery procedure, usually synthetic polymeric macromolecules are used as drug binder that dissolves in suitable solvent. Examples of commonly used synthetic polymers are polyethylene glycol, poly(lactic-co-glycolic acid), and polyanhydride; and examples of natural polymers include collagen and hyaluronic acid (hyaluronan). Most have a high molecular weight and are polydispersed. Some have irregular repeat units and many are branched. All of these things can hinder the best efforts to implement characterization studies. The degradation products of some of the delivery materials do not degrade within a reasonable period of time as required by the application (this may vary from days to months) and some of the degraded products are toxic and create an inflammatory response. Based on the needs of certain applications, the polymer or the drug supportive material should demonstrate versatile mechanical properties (e.g., stent coatings require drug carrier polymers to be elastomeric and microsphere processing requires them to have high glass transition temperature, etc.). It is very difficult for a single drug carrier or polymer to match all of the above criteria. This has led to the development of application-specific nanocarrier that may have the structure-property variability to encompass all potential applications.

To begin this chapter, let us have a look at two fluorescent carbon quantum materials, i.e., zero-dimensional carbon dots (CDs) and two-dimensional graphene QD (GQDs). Both are mainly composed of sp^2 carbon, oxygen and nitrogen elements and other doped heteroatoms [24, 101], which have been widely investigated in the last decade as possible material for drug delivery. However, it must be mentioned here that the size of both of these luminescent carbons varies. The size of CD is <10 nm, while that of GQDs obtains a lateral dimension of up to 100 nm. Therefore, applicability of CD for drug delivery has been found to be more suitable.

4.2 Considerations in Using CD as Drug Delivery Vehicle

One of the main reasons for considering CD as an alternative drug delivery vehicle is that the traditional drug carriers do not have the observability and traceability. The need for nano-size fluorescent nanomaterial CD as drug carrier is expected to solve these problems. The properties of CD have already been discussed in Chapter 1.

Important properties, like biocompatibility of CD, has led to the increasing interest of researchers to use it for various biological applications, especially targeted drug delivery, while also bioimaging it. It serves as an efficient candidate for drug delivery owing to its simple synthesis, environmental friendliness, cost, high quantum yield, abundant source, low cytotoxicity, superior chemical- and photostability and ease of surface functionalization. Recently, CDs have entered the field of drug delivery where it can be speculated that they can act as excellent dual-functional nanocarrier based on the following properties:

- As per the requirement of morphological and other physical properties for particular drug and the target of delivering the payload, CDs can be synthesized using different precursors, dopants and synthesis methods. Most of the methods are eco-friendly and not expensive.
- Mostly CD synthesized by biogenic methods exhibit high quantum yield.
- The particle size of CD can be manipulated for rapid clearance by phagocytic cells, allowing both passive and active targeting. For controlling the size of CD it is important to understand the quantum phenomena. As mentioned in Chapter 1, CDs are also prepared by electrochemical cutting or thermal oxidation of graphite, which gives a mixture of different sizes (such as 1–10 nm) that are separated

sizewise using different methods, e.g., dialysis, chromatography, electrophoretic separation, current density-controlled electrochemical method and density gradient separation by centrifugation.

- Due to the porous nature of CD, drug loading capacity can be easily enhanced by skipping harsh chemical reaction and maintaining the activity of drug.
- The drug release profiles can be altered according to the need by employing polymers and pH-responsive linkers.
- CD could serve as a potential candidate for administration through different routes (oral, pulmonary, nasal, dermal, etc.).
- CD has been found to be able to administer sustained and controlled drug release formulations.
- Moreover, CD can be used as a contrast agent to monitor the drug release of these formulations.
- One of the prerequisites for any bioapplication is negligible cytotoxicity. Umpteen number of investigations at the *in-vitro* level have proved that they are either very low or are completely non-cytotoxic. Comparatively less work has been done on *in-vivo* cytotoxicity studies of CD.

4.3 Designs of CD-Based Drug Delivery System

Designing CD to act as a Trojan horse depends on the drug molecule to be attached according to the functionalization of CD; a ligand is chosen according to the pH of the matrix where drug is to be delivered and the site where drug is to be delivered needs a particular size, e.g., to cross the blood-brain barrier the size of CD and drug conjugate has be <20 nm. Below some examples of consideration when designing CD are given.

4.3.1 Designing for Water-Insoluble Drugs

There were four considerations employed by Jing *et al.* [254] to fabricate (multifunctional core-shell capsules by coaxial electrospray method) CD as drug carrier for the water-soluble anticancer drug paclitaxel using i) olive oil as a reservoir for drug, ii) dual-layer porous titania (TiO_2) shell to suppress the initial burst release of the drug paclitaxel, iii) Fe_3O_4 within the oil core for magnetic targeting, and iv) CD placed within the core functioned not only as drug carrier but also for fluorescence imaging to provide optical information for *in-situ* monitoring of the drug release. This multifunctional core-shell capsules system was helpful in controlled release and delivery of water-insoluble drugs.

4.3.2 Designing for Targeting Tumor Location

One of the problems of delivering drug to the tumor using CD is insufficient uptake of CDs by solid tumors. To combat this problem CD is conjugated with navigating and targeting molecule so that the drug can be delivered to the targeted tumor locations.

A sensitive FRET-based CD has been fabricated for targeted drug delivery via a direct surface coupling. Here CD served as both the donors of FRET pairs and as drug carriers

This system has excellent targeting ability to deliver the drug, as well as very good drug release efficiency to the cancerous tumor cells in acidic pH. Moreover, the capabilities of the conjugate for two-photon imaging and monitoring of drug release in tumor tissues with thickness of 65 ~ 300 μm were demonstrated under the excitation of 810 nm by Tang *et al.* [255]. Mewada *et al.* have synthesized CDs using sorbitol as precursor, surface stabilized with bovine serum albumin and further conjugated with DOX and folic acid [50]. Here, folic acid serves as a targeting molecule for overexpressed folate receptors on certain types of cancer. In this work, the specific internalization of conjugate with and without folic acid has been studied on Vero cells (folate receptor negative) and MCF 7 cells (folate receptor positive) (Figure 4.1).

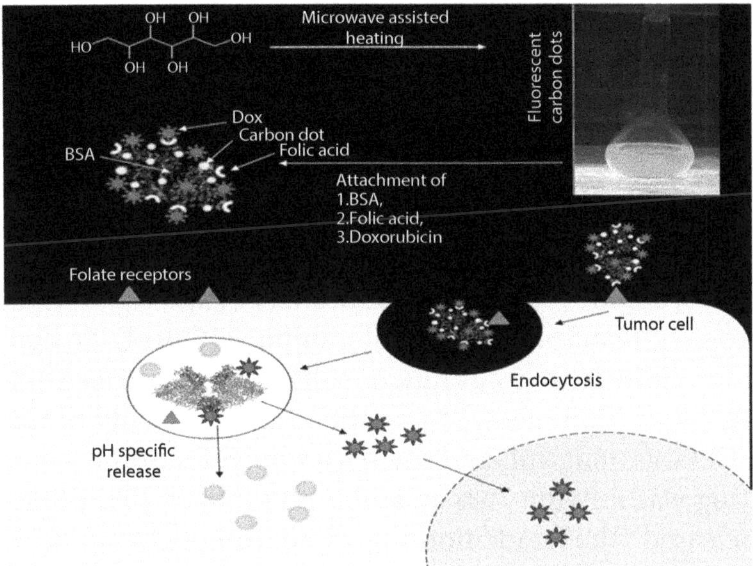

Figure 4.1 Schematic representation of targeting cancer cells using folic acid-conjugated CDs. (Adapted from [50])

4.3.3 Designing a Theranostic Nanomedicine

A theranostic nanomedicine is composed of a therapeutic molecule attached to a nanoparticle. Zheng *et al.* [253] have attempted to synthesize a nanomedicine by conjugating the anti-cancer drug oxaliplatin on the surface of CDs. This conjugate can enter the cancer cells by endocytosis and helps in controlled drug release. Its application has yielded a monotonically decreased tumor size over therapy treatment time. Needless to say, the conjugate possessed bioimaging ability both *in vitro* and *in vivo*, and facilitated the track of the drug molecules by monitoring the fluorescence signal generated by CDs.

4.3.4 Designing a Photoresponsive Nano Drug Delivery System

In photoresponsive drug delivery system intensity of the external light is used as a tool to trigger drug release from the nanocarrier. A photoresponsive nano drug delivery system was constructed by Karthik *et al.* [256] by covalently anchoring a phototrigger conjugated anticancer drug onto the surface of CD. The drug + CD conjugates were readily internalized by the HeLa cells and accumulated within the nucleus and they exhibited precise controlled drug release to kill cancer cells upon irradiation.

There are some phototrigger anticancer drugs, e.g., 7-(3-bromopropoxy)-2 quipolymethyl chlorambucil (Qucbl). To deliver Qucbl to HeLa cells, it was functionalized on the surface of CD. Karthik *et al.* [256] showed that the loading capacity of Qucbl onto CD was 36μg/mL and the release kinetics of the anticancer drug was initially 73% and after 30 minutes 20% of Qucbl was released. The irradiation was done using light (≥ 365 nm, Hg vapor lamp, 120 mW cm^{-2}) and He–Ne laser (5 mW cm^{-2}). This study was carried out using cancerous HeLa cells that were incubated with different concentrations of Qucbl-CDs under irradiation with ≥ 365 nm light for 30 min. The photoreleased Qucbl caused the death of HeLa cells. Moreover, irradiation increases the cytotoxicity for cancer cells treated with Qucbl-CDs compared to those treated with free Qucbl, indicating enhanced photorelease of the anticancer drug inside the cells. The interesting observation was that internalization of Qucbl-CDs were found in both the cytoplasm and nuclei of cancerous HeLa cells.

4.3.5 Designing for Gene Delivery

For gene delivery, usually cationic polymers are used. Such polymers can encapsulate negatively charged DNA. However, the limited serum stability and potential toxicity of cationic polymers limit their applications in gene delivery. Specifically, functionalized CDs can overcome these limitations and act as nanocarriers for delivery of genes.

Liu *et al.* [257] PEI functionalized CDs, prepared by one-step microwave-assisted pyrolysis of glycerol and bPEI25K mixture.

PEI conferred a positive charge on the surface of the CDs and enhanced the fluorescence of the CDs. DNA was attached to the PEI present on the CD surface by electrostatic interactions between DNA and PEI and composite of condensed DNA was formed. The composite was easily internalized by COS-7 cells or HepG2 cells and showed higher gene expression of plasmid DNA than control PEI.

In another attempt, bPEI was used as a carbon source that was oxidized and modified by hydrothermal reaction to get bPEI-based CDs [100]. This CD-bPEI conjugate is highly biocompatible. The transfection experiment demonstrated that the bPEI-based CDs mediated and visualized the gene transfection process in 293T cells and exhibited high transfection efficiency. However, both the transfection efficiency and toxicity of PEI were associated with its molecular weight.

To overcome the molecular weight-related defect, Cheng *et al.* [258] synthesized polycation-bpoly zwitterion copolymer-grafted CD and used it as gene delivery vectors. As compared to PEI25K, the copolymer grafted CD showed superior hemocompatibility and inferior bovine serum albumin protein adsorption. The CDs in the transfected cells kept the excitation-dependent fluorescence property. This block copolymer-modified CDs are promising theranostic vector to deliver genes.

4.3.6 Designing for Antibiotics Delivery

When used as therapeutic drug, antibiotic molecules often result in some side effects. Since rather higher concentrations are administered to the patient in traditional therapy, the side effects are very unacceptable. With the help of targeted nanocarriers it is possible to reduce the drug dosage. Various designs of CD-based antibiotics are being tried to carry antibiotics to the site of action, i.e., the diseased cell of the infecting microbes.

Recently, CD-coated alginate beads (CA-CD) were prepared and used as nanocarriers to deliver tetracycline (TC).

A composite having tetracycline, α-cyclodextrin (TC) and CA-CD were fabricated by electrostatic interactions between the positively charged CD and negatively charged alginate. α-Cyclodextrin is commonly used for attachment of drugs in the central cavity. The TC loading capacity of CACDs was found to be 77% [231]. CA-CD nanocarriers released 70% of TC after 96 hours at pH 1, indicating their applicability in the gastrointestinal tract. The association of α-TC with CA-CDs promoted slow and sustained TC release.

The broad-spectrum antibiotic ciprofloxacin was successfully loaded by Sharon's group [72] onto CDs (Cipro@CDs conjugates) with a loading capacity of approximately 99%. The release profile of ciprofloxacin from CD-based nanocarriers was sustained for 24 hours under physiological conditions. Moreover, Cipro@CDs showed no toxicity toward Vero cells, similar to free ciprofloxacin. More than 90% of Vero cells were viable after treatment with bare CDs at up to 80 mg/mL. The nanocarriers Cipro@CDs displayed potent antibacterial activity against both Gram-positive and Gram-negative strains.

4.4 Carbon Dots for Delivery of Anti-Cancer Drug

Use of CDs in drug delivery system overcome some of the problems associated with drug therapy such as insufficient drug concentration in blood due to poor absorption, less drug solubility and bioavailability and high fluctuation in plasma level.

The advantage of the small size of CD in drug delivery system is that it enables direct entry of entrapped molecules passively (non-specifically) or actively (targeting ligand on surface of molecules). Average human cell diameter is about 20 μm and size of the organelles varies from 20 nm to 800 nm. The size of CDs conjugated with passivating agent, ligand and drug molecules is in range of nanometers; therefore, they directly react with biomolecules on the cell surface or in various mechanisms inside the cell in a non-invasive manner. Another application of CD is it enhances the solubility of the drugs. The increased solubility and bioavailability leads to decreased drug dose, cost and side effects.

4.4.1 A Brief Introduction to Cancer

New cells are created and get replaced with dead cells every moment in our body. The human body is capable of repeating this process a number of times with precision. However, there are possibilities of a small deviation from normal replicating activity of cells due to an intelligent alien influent leading to an unregulated replication of cells and tissues; this may lead to formation of a very large number of cells, leading to tumor formation in the body or formation of deformed cells in the body.

A commonly used scientific definition classifies cancer as a number of diseases in which a healthy cell of the human body gets modified due to genetic defects and starts to grow in an uncontrolled way, invading the adjacent tissues and spreading throughout the whole body as a metastasis. The defects in cells may be caused by chemicals—carcinogens—or may randomly occur due to errors in DNA replication, repair and recombination processes. People of all ages can be affected by cancer and the risk for most types increases with age.

Cancer is not confined to humans; animals and other living organisms can get cancer. When a cell is damaged or altered without repair to its system, the cell usually dies. Also, when such damaged or unrepaired cells do not die they turn into cancerous cells and proliferate with uncontrolled growth, and a mass of cancer cells develop. Frequently, cancer cells can break away from this original mass of cells, travel through the blood and lymph systems, and lodge in other organs where they can again repeat the uncontrolled growth cycle [259]. This process of cancer cells leaving an area and growing in another body area is termed metastatic spread or metastatic disease. For example, if breast cancer cells spread to a bone (or anywhere else), it means that the individual has metastatic breast cancer.

4.4.2 Necessity of Drug Targeting in Cancer Therapy

Destructive methods (e.g., chemotherapy and radiation therapy) have been widely used across the globe as a main cure for cancer. Chemotherapy mainly uses highly toxic drugs to kill cancerous cells; however, along with cancerous cells healthy cells also bear the brunt of the drug. Similarly, use of radiation kills not only cancerous cells but also healthy cells of the body. Such curing methods are undoubtedly saving lives but with unexpected side effects which cause more pain and stress to the patients. Hence, it will be much better if an alternative method is found where cancerous cells can be killed without harming healthy cells around them. With the advent of CD, the possibility of targeted killing of unwanted cells can be achieved. Targeted drug delivery is a special form of delivery system where the pharmacologically active agent is selectively targeted or delivered only to the site of action. The properties for a carrier material to be used for targeted drug delivery should include nontoxic nature, less immunogenicity, physical and chemical stability, biodegradability and should be readily eliminated from the body. The main advantage of targeted drug delivery is that the release rate of the therapeutic agent can be controlled on a particular site with predetermined release kinetics. Surface functionalization of drug carriers with ligands or markers that are selectively recognized by receptors on the surface of cancer cells is an important strategy. Since ligand-receptor interactions are highly selective, it will allow more precise targeting at the site of interest. Specific binding to the cell surface will also enhance endocytosis, thus delivering the drug directly to destroy the nucleus of cancer cells.

4.4.3 Targeting Angiogenesis with CD

Angiogenesis is a physiological process associated with many types of tumors in which the cancerous cells show proliferation of blood vessels around the tumor through which nutrients are supplied for rapid multiplication of tumor cells. Blocking angiogenesis or restricting the growth of blood vessels is considered a promising route for anti-cancer therapy. Shankar *et al.* [260] found that CD prepared from styrene soot displays green luminescence and shows anti-angiogenic effect. Their research showed that addition of CD reduced the angiogenesis activity significantly *in vitro*. It was discovered that addition of CD leads to less expression of cellular growth factors responsible for blood vessel growth. Though the precise mechanism for inhibition of angiogenesis activity is not determined, it is hypothesized that these fluorescent nanoparticles blocked growth factor activity through binding of surface groups on these proteins (Figure 4.2).

4.4.4 Various CD Conjugates for Delivering Anti-Cancer Drug

Chapter 3 of this book focused attention on various CD conjugates that have been designed and fabricated for different applications. Here, we touch on this aspect for delivering anti-cancer drug.

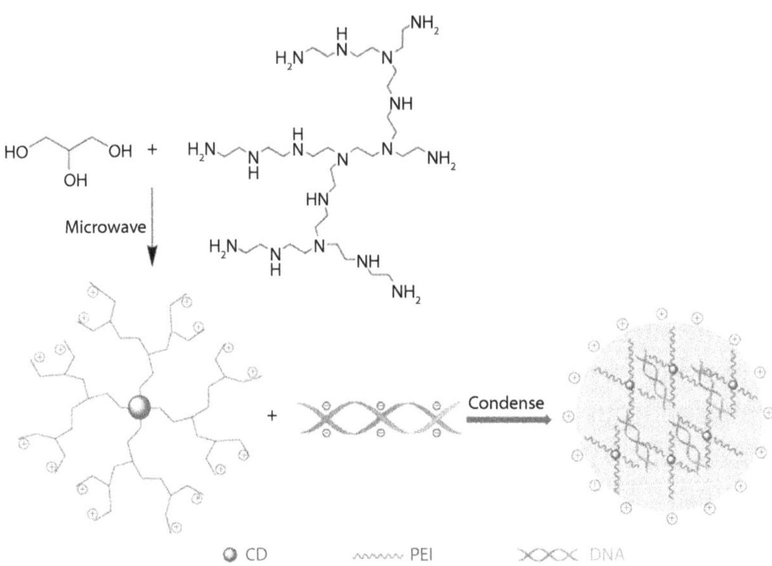

Figure 4.2 Formation of CD-PEI and CD-PEI/pDNA complex. (Adapted with permission from [260])

Lai *et al.* (2012)[182] have synthesized CD from glycerol by pyrolysis and modified the CDs via PEG and controlled the particle size through mesoporous silica nanoparticles. This conjugate of CDs (CDs@mSiO_2-PEG) was uniform and had excellent optical properties. It was used as carrier of the anti-cancer drug doxorubicin (DOX) to HeLa cells and to study the release rate of DOX in HeLa cells. Presence of this conjugate was recorded in the cytoplasm and nucleus where it emitted blue fluorescence and red fluorescence, respectively. The proportions of the two fluorescent colors suggest that this method can be used not only as observable carrier but also for *in-situ* monitoring of release rate of DOX in cells.

4.4.5 CD for pH-Dependent Drug Release

The changes in pH between healthy and diseased tissue have been exploited to design pH-triggered conjugated drug delivery system for release of active drug at selected site. CDs have been experimented on to conjugate with other nanomaterials owing to their surface properties and excellent fluorescence capability. Zhao *et al.* [261] have reported utilization of mesoporous silica nanoparticles for delivery of DOX where CD synthesized using PEI have been attached on the pores and act as gatekeepers for the release of model drug. Further hyaluronic acid was attached to obtain controlled drug release. The results suggested that the whole complex exhibited enzyme and redox dual-responsive drug release. Also, excellent fluorescence property and biocompatibility were observed due to the presence of CD on the surface.

In another report, Sharon's group [184] have synthesized CD@mesoporous silica complex using rice husk by one-pot synthesis. The synthesized complex displayed green fluorescence under UV light and has been tested as a bioimaging material using yeast cells. One-pot simultaneous synthesis of the complex reduced the extra attachment step and drug can be directly loaded. The porous nature of silica nanoparticles would serve high drug loading efficiency along with CD acting as nontoxic imaging agent. In an interesting work, CD conjugates have been utilized for photodynamic therapy (PDT). PDT is based on singlet oxygen releasing photosensitizer on tumor cells. The oxygen radical is released by irradiation of Ps using light. PDT can be employed using CD based on its two-photon absorption property. Chen and workers designed CD covalently attached to porphyrin derivatives [262]. These porphyrin derivates absorb light in the visible range and hence cannot be applied for therapeutic purpose directly. However, the designed CD absorbed light in NIR region, which penetrates through the body tissues. A shorter wavelength of light is emitted by CD through a two-photon process, which leads to the excitation of the porphyrin derivatives. This combination of NIR light absorption and fluorescence resonance energy transfer (FRET) lead to killing target cell. Such work suggests that dual-stimuli responsive systems along with real-time imaging can be explored more using CD with other nanomaterials and thus opens a wide area for research.

4.4.6 CD for Drug Delivery to Renal Cancer

Renal or kidney cancer arises when healthy kidney cells grow out of control, forming a mass called renal cortical tumor. This tumor could be malignant (cancerous), indolent (cancerous but rarely metastasizes), or benign (non-cancerous).

Though CD has excellent fluorescent property, high photostability against photobleaching and blinking, broad excitation spectra, narrow and tunable emission spectra, good biocompatibility, small size and low toxicity, so far very little attention has been paid to its use in delivering drugs to renal cancer cells. The small size of CD (i.e., in the range of 2–10 nm) that possesses sp^2/sp^3 disordered carbon or hybridized graphitic core functionalized with polar carboxyl or hydroxyl groups on the surface are the favorable properties of CD that will help in its easy excretion from the renal system after the delivery of the drug. Apart from the above-mentioned properties, amalgamation with their broad absorption spectra, good biocompatibility, large absorption coefficients, and good solubility in polar solvents make the CDs a very attractive possibility for its use in delivering drug to renal cancer carcinoma cells. The ability of CD to exhibit PL emission in the near-infrared (NIR) spectral region under NIR light excitation is particularly significant and useful for *in-vivo* studies because of the transparency of body tissues in the NIR "water window." Interestingly, the PL from CDs can be quenched efficiently by either electron acceptor or electron donor molecules in solution, indicating that photoexcited CDs are excellent electron donors and electron acceptors. Because of its photoluminescent property, CD can be used to track fundamental molecular events occurring in renal tumor cells. There is a need to research this possibility. Since renal cell carcinoma is a multifactorial disease the complications that arise are i) development of resistance and ii) toxicity due to non-specificity of drug delivery. Therefore, renal cancer therapy needs either single-drug inhibitor or multi-drug inhibitors. The advantage in using CD will be that it can have functionalized sites to attach multi-drug inhibitors.

4.4.7 CD for Drug Delivery to Lung Cancer

Though surgery is the first option for localized lung cancer, it may lead to relapse and hence the patient may have to opt for other treatments like chemotherapy, radiation therapy or sometimes even a combination of both. However, side effects related to these treatments open up a scope of research in the field of nanotechnology where, in the near future, aerosol-like drugs can be used which can be inhaled and provided to the active site to reduce pain and increase patient compliance.

Hollow CDs have been used to deliver doxorubicin (DOX) [263]. The release of DOX was found to be high at acidic pH. Cellular uptake studies of DOX-CD complex showed that CDs were localized in the cytoplasm of A549 cells. In contrast, DOX was observed in the nucleus of the cell. The authors proposed that the complex might have been internalized by endocytic vesicles and further transported to the lysosome. Due to the acidic condition of lysosome, DOX gets released from the CD and enters the nuclei causing cell death. It was also observed that DOX-CD complex was internalized faster as compared to free DOX due to possible different mechanisms.

4.4.8 CD for Drug Delivery to Breast Cancer

Breast cancer has been growing at a fastest rate than any other cancer among women worldwide. Emerging trends in nanotechnology have opened great opportunities to design nanoparticle-based therapeutic agents. Wang and coworkers [264] have recently detected the presence of fluorescent CD in commercial beer with average size of 2.5 nm. The CDs showed high solubility, quantum yield and were found to be amorphous in nature. The authors studied the cytotoxicity analysis of DOX attached CD on MCF-7 cells. DOX-conjugated CD displayed prolonged cytotoxicity as compared to free DOX due to sustained release mechanism. In another report, fluorescent nitrogen-doped CD synthesized using dried shrimp was used as drug delivery vehicle for targeted delivery of boldine in MCF-7 cells. The internalizing vesicles trigger the release of boldine from the CD due to pH difference, thus allowing the anti-cancer drug to kill MCF-7 cells. pH-Mediated release of drug and its diffusion to the target site (nucleus) was found to be beneficial in the case of boldine. At acidic pH the percentage of drug release shoots up with increasing time (maximum value of 84.8% after 36 hours) and the rate was found to be higher than that of physiological pH. Cellular uptake studies revealed that bare CD is easily accumulated in the cytoplasm of MCF-7 cells.

4.5 CD for Drug Delivery to Neurodegenerative Disease

The central nervous system represents a formidable challenge for the delivery of therapeutic agents due to the blood-brain barrier (BBB). This physical barrier limits the brain uptake of the vast majority of therapeutic and imaging agents. Substances may gain access to the central nervous system by lipid-mediated free diffusion or potentially by receptor-mediated endocytosis of nanoparticles. The most pivotal role of CDs, due to their small size, is that they can cross the biological barrier and extravasate through blood and tissues. This property is very important in tumor vessel, which is often dilated or fenestrated with average pore size average less than a micron [265]. Nanotechnology may provide an effective means for circumventing this delivery issue past the BBB. Nanoparticles may be designed to mimic low density lipoprotein (LDL) and interact with the LDL receptor, thus causing uptake by endothelial cells via receptor-mediated endocytosis. Nanoparticles may effectively mask covalently bound chelators, thus facilitating their delivery past the BBB, and minimize toxicity while improving the pharmacokinetic effect of the chelator itself. To facilitate transport through the BBB via the LDL mechanism, the nanoparticles may be further functionalized with apolipoproteins.

Sharon *et al.* have given one of the initial reports on the synthesis of CD from phenylalanine by microwave-assisted method and delivery of the antipsychotic drug haloperidol *in vitro*. The drug covalently linked onto the surface of CD via cysteamine HCl linker. The drug release study suggests controlled release of haloperidol at physiological pH. This system can act as a potential vehicle for delivering haloperidol for treatment of schizophrenia [71].

4.6 CD for Gene Therapy

The field of gene transfer/therapy has successfully employed CDs. The initial study by Liu *et al.* [188] has shown use of polyethylenimine (PEI)-modified CD to absorb negatively charged DNA for gene transfer. The researchers found that CD could not only successfully transfer the plasmid DNA, but also retained their multicolor fluorescence property during the process upon multi-wavelength excitation. The DNA-PEI-CD composite was internalized by COS-7 and HepG2 cells, which showed higher gene expression of plasmid DNA than control PEI. Wang *et al.* [266] have prepared hybrid gel using a mixture of chitosan, PEG and CD which displayed NIR light/pH dual response. Similarly, bPEI was used as a carbon source for the preparation of CD using hydrothermal method which showed excellent biocompatibility.

In order to overcome the inherent molecular weight-related toxicity of PEI, Cheng and his coworkers [258] synthesized polycation-b-polyzwitterion copolymer-grafted CD and used it as an efficient delivery vector. These CDs displayed condensation of plasmid DNA on their surface with approximate hydrodynamic diameter of 50 nm. Greater transfection efficiency and lower toxicity were observed at high ratios. Thus, block copolymer-modified CD can serve as promising therapeutic vectors to deliver genes. A few other studies have demonstrated direct synthesis of CD from oligonucleotides as carbon source. Ding *et al.* [153] synthesized fluorescent CD hydrothermally using purified natural DNA from *E. coli*. It was observed that DNA-extracted CD could be readily internalized into the cells. These CDs have been successfully employed to deliver different drugs into bacterial and mammalian cells. Owing to its readiness to cross the cell membrane, such CD can be used as a potential platform for transporting genes into the cells.

4.7 CD to Monitor Delivery of SiRNA

SiRNA technology is an emerging field which employs small interfering RNA fragments to turn off expression of target genes. The biggest challenge lies in transporting the negatively charged SiRNA across the cell membrane, thus restricting its complete practical application. CDs have been successfully used as scaffolds for intracellular ferrying of such molecular cargos. Cui and colleagues [267] used CD as a delivery vehicle for SiRNA. The CDs were coated with positively charged PEI which served as a docking site for negatively charged SiRNA through electrostatic attraction. PEI served a dual role in this research. Firstly, it enabled the transport and uptake of CD complex across the membrane and into the cell. Secondly, it functioned as a carrier for SiRNA molecules by attachment on positively charged CD surface. The most important part of this research was that the whole complex retained an overall positive charge, which is a prerequisite for transport through cell membrane. Similarly, PEI-functionalized CDs were successfully used to carry and deliver Survivin-SiRNA into human gastric cell line MGC-803 [158]. An important advantage of using CD as intracellular delivery vehicle is its combined role to transport the molecular cargo as well as enable imaging due to its fluorescence properties, allowing assessment of the effectiveness of the process.

4.8 Challenges in Using CD as Drug Delivery Vehicle

There are some successful applications of nanotechnology in drug delivery, as evidenced by a few nano-drug products on the market. But not all approaches have been successful. Use of new nanomaterials, such as CDs, come with challenges which are being surmounted by tackling modification of the physicochemical characteristics of the CDs to improve the required properties; for example, i) long circulation period in the blood,

ii) increased functional surface area, iii) protection of incorporated drug from degradation, iv) capability to cross the biological barriers, and v) site-specific targeting.

Another challenge is scaling up the research and development (R&D) outcome for large-scale production and eventual commercialization because, despite the large number of patents, commercialization of nano-drug delivery system is still at large. Scaling up challenges need attention for including the low concentration of CD, though CDs do not agglomerate as other metal nanoparticles. Still, this consideration has to be looked into along with the chemical process and maintaining the size and composition. Another unfortunate reason for the delayed commercialization is that many pharmaceutical industries are yet to consider this new technology, which may be due to the lack of regulatory guidelines that would account for physicochemical and pharmacokinetic properties of nano-drug products, which are different from conventional drug products, and also the challenges of scaling up. Conventional safety measures in a pharmaceutical factory may not be appropriate for the production of nano-size materials. One more challenge that needs attention is the safety issue during R&D, scaling up and commercialization. Attention should be paid to the biodegradation and the side effects of the by-products; also, materials should be safe at macroscale since there may be a change in physicochemical characteristics at nanoscale, and nanocarriers should be totally cleared from the body, as their accumulation may have several possible effects.

4.8.1 Prevention of Drug from Biological Degradation

The plasma binding profile on any drug delivery vehicle contributes to its longevity in the blood stream. This will have important consequences on its efficacy [268]. Any change in hydrodynamic diameter can be studied by incubating the CD conjugated with drug in human plasma. An approximate doubling or increase in diameter would mean plasma binding is positive, and hence would prevent any fast degradation of drug till the vehicle reaches the target site. Moreover, closely related plasma-like proteins can be used for coating the CD drug conjugate, thus decreasing its ability to be degraded before reaching the target. The plasma-like proteins can be engineered to break down with a pH change or more specificities for release of drug at the proper site in the body.

4.8.2 Effective Targeting

Any targeting molecule can be attached easily due to its broad-spectrum surface properties and the multiple functional groups. Many types of cancer overexpress markers or antigens against which specific ligands or antibodies can be used for targeting. Mewada *et al.* [50] have attached folic acid (FA) onto the surface of DOX-attached CD. Many types of cancers overexpress folate receptors on their cell surface which could be directly used for delivering drugs to specific sites. This research group has worked with MCF-7 cells which overexpress folate receptors. The targeting was studied by incubating MCF-7 and Vero cells with CD-DOX-FA complex, as well as bare CD. CDs were internalized in both the cells, as could be visualized using fluorescence microscopy. It was clearly observed that uptake of FA-decorated CD was found to be greater in MCF-7 cells as compared to Vero cells due to the presence of receptors on the former cells.

Wu *et al.* [219] prepared CD which had a dual function of targeting and delivery along with imaging. The first prong was mitochondria recognition moiety (triphenylphosphonium, TPP) and the second was nitric oxide (NO)-releasing moiety. The NO acts as a potent anti-cancer agent by blocking the mitochondrial respiration. The researchers showed that dual-pronged CD was bound to mictochondria in the cell and released NO, thus destroying the cell through apoptosis.

4.8.3 Patient Compliance

Inherent biocompatibility of CD renders them as potential candidates for theranostics. One of the major concerns about the patient population is the safety and possible impact of CD as a carrier for drug and genes; this concern also includes the entire manufacturing and disposal process. Measures are needed to take extra care of the environment and protect it from the negative impacts of nanomaterials.

4.8.4 Cost Effectiveness

It is envisaged that the cost of nanotechnology-based therapeutic system will be much reduced, since less amount of materials are expected to be used for production as compared to the production of conventional therapeutic materials. However, pharmaceutical companies are doubtful about it because they will be considering the patent and excessive R&D cost involved that they would like to recover for successful commercialization. So, it may turn out to be expensive. The general public is reluctant to embrace nanotechnology-based product, as was the case with genetically modified products. However, despite these challenges, CD-based drug delivery is a development that cannot be ignored and so the challenges are being tackled.

Many natural precursors can be used to very easily synthesize CD. There are various simple, eco-friendly, low-cost synthesis methods of fluorescent CD [269]. This versatility in synthesis methods of CD reduces the overall cost of production. Various researchers have attempted synthesis of CD from different low-cost products as well as waste products like kitchen wastes, fruit peel [270], willow bark [271], jaggery [269], polysaccharide gum [72], sugarcane juice [66], etc. Moreover, CD synthesized from natural sources gets inherently surface passivated with different functional groups which help in the direct attachment of drugs. No harsh chemicals are required for reduction of carbon compounds derived from the above-mentioned precursors.

4.9 Suitability of CD-Conjugated Drugs

Conventional drug delivery technique has already advanced drug delivery to patients through various routes. Taking a clue from them and with innovative ideas, nanotechnologists have already started working in this direction. These techniques are discussed in brief below.

4.9.1 For Oral Drug Delivery

Generally, polymer complexed nanoparticles or polymer-based nanoparticles are preferred for oral delivery of drugs to the stomach, intestine, etc. The water uptake, swelling and release of encapsulated drug can be modified by adjusting the porosity of nanoparticles during their synthesis. Alternatively, different densities of the same polymer or different types of polymers can be used to achieve variable drug release. CD-coated alginate beads have been successfully prepared for the delivery of tetracycline with β-cyclodextrin [220]. CD coating enhanced not only the stability and swelling efficiency of beads but also increased the antibiotic loading capacity. Sustained release of antibiotics at pH 1 was recorded using *in-vitro* system. The results suggest that such CD polymer beads complex can serve as a potential platform for delivery to the gastrointestinal tract (pH is low), which would give long-time release of drug after administration.

4.9.2 By Inhalation

Drugs delivered through inhalation are generally in suspension form. CD being small in size and partially porous in nature ensures high drug loading efficiencies [239]. Easy synthesis of water-soluble CD also makes highly soluble water suspensions. Furthermore, the rate of release of drug can be pre-decided by using specific linkers, which will offer sustained release for longer periods of time. In the case of drug delivery to the lungs, an accurate amount of drug reaching the target can be ensured.

Thus, such properties make them an ideal candidate for drug delivery via inhalation route.

4.9.3 As Transdermal Drug Delivery

The transdermal delivery system of medicine can be an alternative method of oral, inhalation, direct injection into the blood or other forms of drug delivery systems. Medicine is transferred through a drug reservoir attached on the skin and its rate is controlled by a membrane. The advantage of such a system is that the patient does not need to remember the time to take the medicine. Various nanoparticles and polymers have been used in transdermal delivery mechanism [272]. Due to their ultra nano size and ability to functionalize multiple cargos on their surface, CDs can be used in the transdermal delivery system of medicines. Patra *et al.* [273] have presented transdermal delivery of curcumin via nanosized liposomes. They have further studied that incorporation of CD in these curcumin-loaded liposomes improved their stability as well as permeability through the skin.

4.9.4 As Injection

This is the most important route in the case of delivering drugs attached to nanoparticles. Because CDs are in the size range of 5–20 nm they can easily pass through the smallest blood vessels. Moreover, it is especially preferable for cancer therapeutic drugs to be injected intravenously so that the drug-loaded vehicles easily reach the target site via blood circulation.

4.10 Summary

This chapter has provided insight into the increasing awareness and research being done in the area of CD-based drug delivery system. It is expected to continue to change the whole concept of medicines, including aspects such as product characteristics, bioavailability, pharmacokinetics, stability, drug use, and toxicity in humans.

5

Carbon Dots for Cell Imaging and Diagnostics

"A careful analysis of the process of observation in atomic physics has shown that the subatomic physics have no meaning as isolated entities, but can only be understood as interconnection between the preparation of an experiment and the subsequent measurement."
Erwin Schrodinger

5.1 Introduction

Not only CDs, but many carbon allotropes showing graphitic structure, i.e., honeycomb-like arrangement of carbon atoms, have shown potential for various applications in electronic devices in biosensors and bioimaging agents. Some examples of these are 0D fullerene [274], diamond nanocrystals [123] and carbon dots (CDs), as well as 1D CNT [275], 2D graphene QD

(GQDs) [276] and graphene [277]. However, it must be mentioned here that although CDs and GQDs have similar quantum-confined fluorescent carbon materials, CDs have been more used as biosensors and bioimaging agents [278]. Both CD and GQD are mainly composed of sp^2 carbon, oxygen and nitrogen elements and other doped heteroatoms [24, 101]; the different spatial arrangements of carbon atoms exhibit distinctive physical and chemical properties [279]. Unlike GQD, the CDs do not have perfect crystal structures [42]. Moreover, CDs exhibit luminance at size < 10 nm, whereas GQDs have size up to 100 nm [280, 281] and are luminescent. Generally, CDs display strong optical absorption in the UV region, with a tail extending out into the visible range (Figure 5.1a), i.e., 280–360 nm. The absorption band could be regulated via surface passivation or modification. CDs show the excitation-dependent emission properties in wavelength and intensity (Figure 5.1b). Other properties that make CD more applicable for biomedical applications are their low cost, high quantum yield, abundant source, low cytotoxicity, and superior chemical and photo stability. Wang *et al.* [282] have evaluated the cytotoxicity of the CDs prepared by different combinations of precursor of CDs and molecules used for surface functionalization. It is found that in terms of their nanoscale structure and configuration CDs are

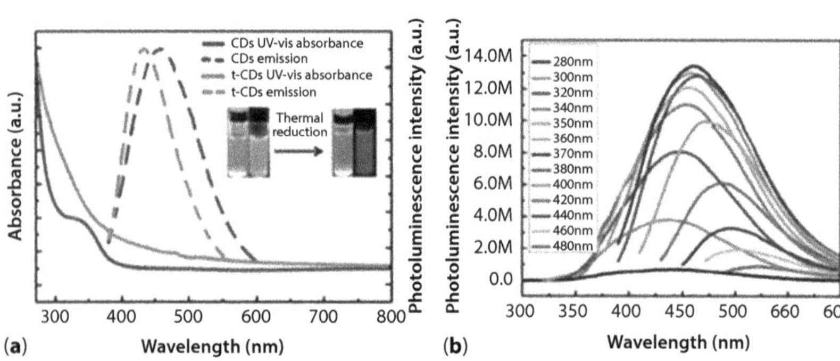

Figure 5.1 (a) UV–Vis absorption and normalized photoluminescence spectra of the CDs and t-CDs. Inset: photograph of the CD (left) and t-CD (right) aqueous solutions under visible light and 365 nm UV light, respectively. (b) Emission spectra of the CDs with excitation at different wavelengths. (Reproduced from [283])

not intrinsically cytotoxic. The cytotoxicity of CD depends on the selection of surface passivation molecules.

The distinct light emission or optical properties of CDs and their tunable band gap is highly impacted and can be controlled by synthesizing the CD from different precursors, methods, functionalization, doping and surface orchestration of CDs. Such variations in the type of CDs have displayed its applicability in the biomedical field, such as for analytical sensors and bioimaging devices.

As mentioned in Chapter 1, CDs have been synthesized from SWCNT by arc-discharge methods [76], electrochemical method using different precursors such as graphite [213, 33], by supported routes [76], combustion/heating [284], hydrothermal synthesis [277], acidic oxidation [92], microwave/ultrasonic method [50], laser ablation [56]), and plasma treatment [285]. Apart from pure carbonaceous materials, such as CNT, graphene or graphites, some other materials like i) Molecular compounds such as ammonium citrate, octadecylammonium citrate, sodium 11-aminoundecanoate, etc.; ii) Small organic compounds such as citric acid, boric acid, amino acid, glycerol, glucose, ethylene glycol, N-acetylcysteine, diamine and benzene; iii) Chemicals from natural sources, such as orange juice, banana juice, soy milk, meat, milk, coffee, beer, egg, potato, pomegranate fruit, *Bombyx mori* silk, sugar, bread, jaggery, lysozyme, sucrose, starch, grass and peels of water chestnut, have been used as precursor to fabricate CDs. CDs synthesized from such compounds are rich in the presence of functional groups, such as carboxyl, amino, hydroxyl, etc., on the surface of CD, which facilitate further modifications so as to tune the optical properties. The synthetic methods of CDs have been systematically discussed in Chapter 1. Three major requirements for the preparation of fluorescent CDs with low background are: 1) Improved optical properties of CDs by tuning carbon sources; 2) Appropriate passivated agents to enhance the fluorescence of CDs; and 3) Introduction of heteroatoms to change the band gap spacing of CDs to regulate fluorescence yield and emission range. Different synthesis procedures have yielded CD of different properties and that is why the emphasis has been placed on different methods using different precursors as well as addition of other moieties to CD, so that it can be used for a variety of applications.

5.2 Bioimaging

Bioimaging is a method that non-invasively visualizes biological processes in real time without much interference in the living processes. Moreover, it gives information on the 3D structure of the specimen from the outside, i.e., without physical interference. Through bioimaging, subcellular structures and entire cells of tissues up to entire multicellular organisms can be observed. At the cellular level, bioimaging can be used to follow cellular processes, quantify ion or metabolite levels and measure live interaction of molecules at the site where it happens. Appropriate tracers, such as specific fluorochromes, and advanced microscopic instruments such as, for example, confocal laser scanning microscopes (CLSM), are a prerequisite for most applications. Bioimaging is achieved by using light, fluorescence, electrons, ultrasound, X-ray, magnetic resonance and positrons as sources for imaging.

The tunable optical properties and cytotoxicity play important roles in application of CD in biosensors and as bioimaging agent. Initially, semiconducting QD of CdSe and other related core-shell nanoparticles were tried for *in-vitro* and *in-vivo* bioimaging. But due to their high cytotoxicity and environmental concerns, their applications in biosystem and health care remained limited. As mentioned above and in Chapter 2, the special properties of CD, i.e., superior photostability, non-blinking, low cytotoxicity, optical properties, visible excitation and emission wavelengths, and high brightness at the individual dot level, were found to be suitable for both *in-vitro* and *in-vivo* fluorescent probes for imaging of both animal and plant cells. Apart from bioimaging or labeling the cells, CDs have opened a new vista, i.e., their applications in imaging-guided therapy.

Maximum studies have been done on utilization of CDs for bioimaging various cancerous cells both *in vitro* and *in vivo*. Especially after the first work of Sun's research group, who applied CDs in bioimaging of cultured MCF-7 cells and showed their presence mainly in the cell membrane and cytoplasm. After that, there was a flood of work done in bioimaging cancer cells as well as bioimaging in various other cell lines such as virus, bacteria, fungus, plants, mice and humans. For these applications, CDs were synthesized from different precursors and modified, doped and/or functionalized as needed.

The effect of nitrogen doping of CD on *in-vitro* and *in-vivo* bioimaging was investigated by Qian *et al.* [286]. To dope CDs with nitrogen they synthesized CDs by one-pot hydrothermal method using different sources of nitrogen. The nitrogen-doped CDs emitted bright blue fluorescence and their QYs were elevated owing to the CDs being doped with nitrogen atoms (up to 36.3% emission efficiency). They suggested that the enhanced fluorescence may be due to the presence of poly-aromatic structures induced by nitrogen incorporation into CDs.

The important parameters that influence *in-vivo* imaging are:

- **Particle size of CDs:** CDs with hydrodynamic diameter <5.5 nm can be easily filtered through the glomerulus, thus easily get excreted into urine.
- **Surface charge of CDs:** CDs have neutral surface charge, which prevents the adsorption of plasma proteins. Therefore, the chance of accumulation of CDs in the liver or spleen is efficiently reduced, thus accelerating the clearance of CDs from the body [287].
- **Low cytotoxicity and biocompatibility of CDs:** In an interesting study, Yang *et al.* [137] injected (both subcutaneous and intravenous) CDs and ZnS-doped CDs, which were PEGylated with PEG1500N to the backs of mice (subcutaneous injection). They were used for *in-vivo* optical imaging of mice. Both CDs

and ZnS-doped CDs emitted strong fluorescence *in vivo*. The subcutaneously injected CDs showed slow diffusion and accumulated in the liver. The intravenously injected CDs were excreted via urine. Hence, they might be suitable as an important contrast agent. Blood test and histological analysis of CDs done by Tao *et al.* [288] and Nurunnabu *et al.* [289] demonstrated that no toxic effects of CDs-dots to the treated mice were observed.

- **High fluorescence:** One of the limitations of CDs is the relatively low luminescence efficiency as compared to other QDs. However, many biogenically synthesized CDs have shown enhanced luminescence, e.g., CD derived from bagasse by hydrothermal treatment under alkaline conditions exhibited a 20-fold enhancement in luminescence, as reported by Liu *et al.* [290]. Deng *et al.* [291] have reported that the blue emission from CDs depends on the presence of polar groups, i.e., the interactions between C=O bonds on CDs conjugated with aromatic carbons, probably via dipole–dipole interactions or hydrogen bonding between C=O bonds and polar groups. They interpreted that the interactions may modify the electronic structure of C=O bonds and enhance the absorption and emission efficiency of CDs (Figure 5.2). Based on this mechanism one can explain the importance of matrix or surface modifications to promote CD emissions.
- **Sufficient uptake of CDs by cells:** Surface passivation with PEG improves the hydrophilicity of the CDs, influencing their uptake as well as clearance *in vivo*.
- **Ability of CDs to cross blood-brain barrier (BBB):** Lu *et al.* [292] synthesized nitrogen-doped CDs (N-CDs) using a one-pot hydrothermal method. They used citric acid (as CD precursor)

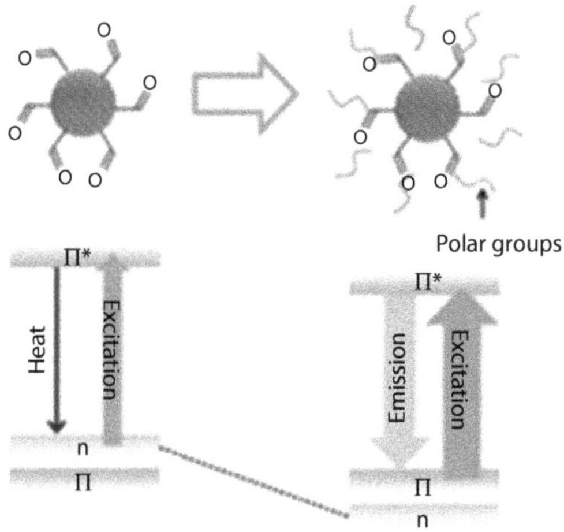

Figure 5.2 Proposed mechanism of environment-dependent blue emission of CDs (Reproduced with permission from [291]; Copyright © 2014 The Royal Society of Chemistry)

and polyethylenimine as passivating agent to get monodispersed and quasi-spherical N-CDs with an average size of ~2.6 nm that emitted a strong blue luminescence with a quantum yield as high as 51% and excellent photostability. They suggested that N-CDs could be applied as efficient nanoprobes for real-time live imaging of live cells. Moreover, due to their small size and the surface positive charge rendered by the PEI, they showed that N-CDs crossed the blood-brain barrier (BBB) in a concentration-dependent manner using an *in-vitro* BBB model, making them potentially useful in traceable drug delivery in brain disorders.

In-vitro imaging provides information on the imaging ability, distribution and cytotoxicity of probes in cells. The important parameter for *in-vitro* bioimaging encompasses the use of upconversion fluorescence. Upconversion fluorescence has a lot of advantages, including the fact that it is noninvasive and has deep penetration of NIR radiation; hence, CDs are more advantageous and are widely researched for bioapplications.

For example, PEI-CDs that exhibit excellent upconversion PL properties with emissions in the range of 308–550 nm by long-wavelength light excitation have been successfully used for *in-vitro* bioimaging with two-photon excitation [116]. CDs show a strong emission with either the 458 nm excitation or 800 nm two-photon excitation. Based on this concept various types of cells have been labeled and imaged *in vitro*.

Fluorescent CDs were directly employed for labeling antibodies via amine-amine conjugations. The sandwich immunoassay approach was utilized to determine human α-fetoprotein (AFP) with capture and label anti-AFP pairs. CDs provided highly stable fluorescence signal corresponding to the human AFP target. Hence, CDs can be applied as bio-labels for high-throughput well-plate-based immunoassays [159].

5.2.1 Bioimaging of Cancerous Cells

For bioimaging or cellular uptake of CDs it is essential that they are functionalized and compatible. A variety of carbon precursors, such as graphene sheets, sucrose, amino acids, chitosan, and biomaterials, and different surface passivation agents, such as glutathione, and polymers like PLGA, PEG, and PEI have been used for the synthesis of carbon dots or structurally and for surface functionalization. CDs have great potential to be used for imaging/labeling cancer cells. Recently, MWCNT-derived CDs were used for *in-vivo* fluorescence imaging of a nude mouse [288]. The mouse was subcutaneously administered with CDs at three different locations (Fig. 5.3a). Blue, green, yellow, orange, red, deep red, and NIR light excitation with center wavelengths at 455, 523, 595, 605, 635, 661, and 704 nm, respectively, were applied during *in-vivo* imaging of the mouse. As explained by the authors, "After spectral unmixing to differentiate the background autofluorescence (green), the subcutaneously injected spots (red) on the mouse were seen in these fluorescence images under all of the different light excitations (Figure 5.3). Compared to images acquired using blue

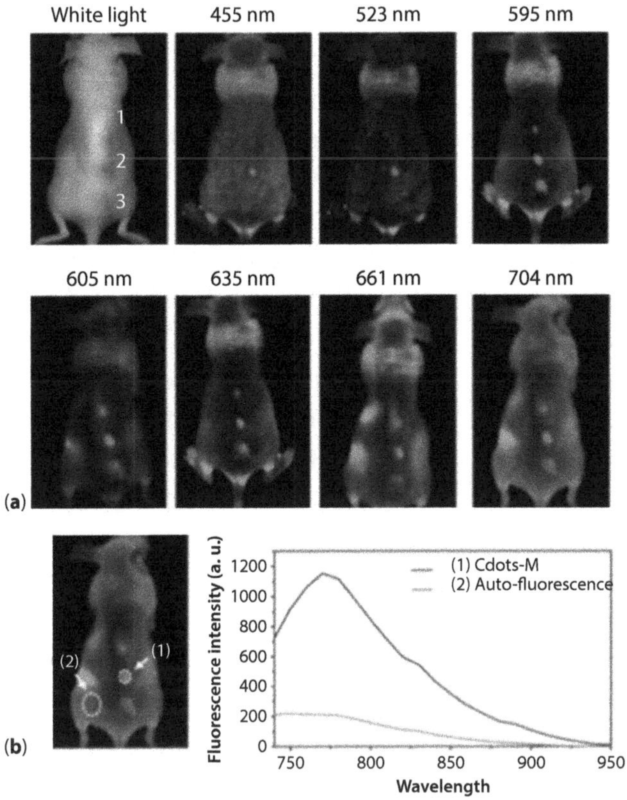

Figure 5.3 *In-vivo* fluorescence imaging. (a) In vivo fluorescence images of a CDs-injected mouse. The images were taken under various excitation wave length at 455, 523, 595, 605, 635, 661, and 704 nm. Red and green represent the fluorescent signals of CDs and the tissue autofluorescence, respectively. (b) Signal-to-background separation of the spectral image taken under the NIR (704 nm) excitation. The fluorescence of CDs was well separated from the tissue autofluorescence background. (Reproduced with permission from [288]; Copyright © 2012 John Wiley and Sons)

(455 nm) and green (523 nm) light excitation, those taken under longer wavelength excitations (595 nm and beyond) showed much better signal-to-background separation. Although the fluorescence emission of CD is weaker at longer wavelengths, the tissue autofluorescence background decreased even more, resulting in an improved signal-to-noise ratio under red and NIR excitation. *In-vivo* optical imaging at longer wavelengths is usually preferred owing to the improved photon tissue

penetration and reduced background autofluorescence, especially in the NIR region. The capability of CDs for NIR *in-vivo* fluorescence imaging (excitation ¼ 704 nm, emission ¼ 770 nm) demonstrates the great potential of CDs for use as optical nanoprobes in biomedical imaging."

For bioimaging, it is essential that CDs are taken up or internalized by the cells in consideration. Xiao *et al.* [293] have suggested a potential pathway for the cellular uptake of CD. According to them it consists of three steps: 1) Endocytosis through phagocytosis and pinocytosis mechanism. Phagocytosis is the uptake of large particles by specialized mammalian cells such as macrophages, monocytes, and neutrophils; whereas, pinocytosis is the uptake of small particles, solutes and fluid by any cell type. Receptors are also involved in endocytosis. Endocytosis is probably assisted by receptors (present on the cell membrane) specific to ligands with negative charges; 2) Sequestering in early endosomes of larger particles is the next desired step; followed by 3) Translocation to later endosomes or lysosomes.

Some successful examples are discussed below.

5.2.1.1 HeLa Cells

HeLa cells are an immortal cancerous cell line. This cell line was derived from cervical cancer *cells.* It is extensively used in scientific research.

For bioimaging of HeLa cells, Sharon's group synthesized CDs by microwave-assisted thermal oxidation using sorbitol as a precursor and NaOH as passivating agent. The conjugate was functionalized with bovine serum albumin (BSA) and then folic acid (FA) was attached as navigating molecule to HeLa cells [50]. The results are presented below (Figure 5.4).

Transferrin receptor (a transmembrane glycoprotein) is overexpressed in tumor cells because of the increased demand of iron; hence, transferrin receptors show higher uptake of CDs. Therefore, when CDs functionalized with the glycoprotein transferrin (having specific binding of ligand and receptor) were used

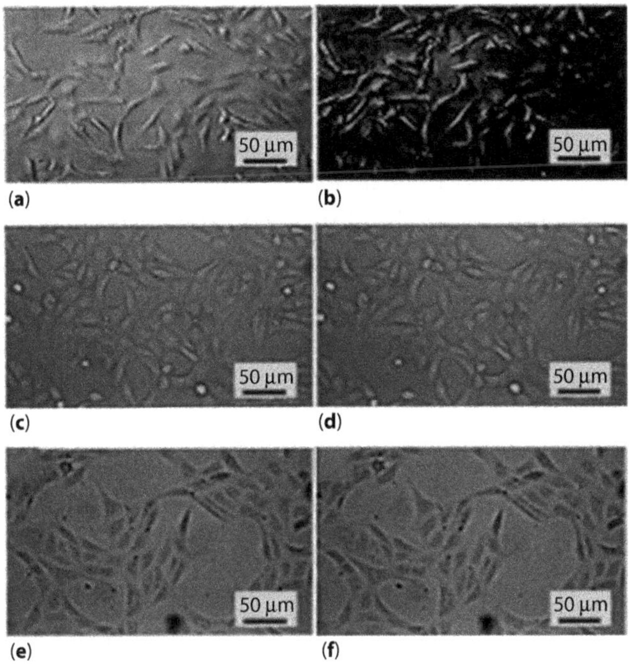

Figure 5.4 Images of cells taken in (left panel) Bright field and (right panel) epifluorescence microscope. (a-b) HeLa cells treated with functionalized CD with FA complex; (c-d) HeLa cells treated with only functionalized CD-dots (without FA); (e-f) Vero cells with functionalized CD-FA complex. (Adapted with permission from [50])

it enhanced the entry of CDs to HeLa human cervical cell line as compared to un-functionalized CDs. Based on these results, Li *et al.* [115] suggested that transferrin-functionalized CDs can be used as nanoprobes for disease diagnosis as well as drug targeting. CDs synthesized by solvothermal technique using glucose as carbon source and modified with PEI to improve the fluorescence of CDs were conjugated with CEA8 antibody for labeling HeLa cells *in vitro*; they did not exhibit any observed cytotoxicity [217]. Moreover, there is speculation that CDs can enter into HeLa cells through non-endocytic pathways [294].

When Ding *et al.* [295] exposed labeled cells to surface modified CD, their images showed that the CDs were gradually taken up by HeLa cells. The LC_{50} lethal concentration (LC50) of CD for HeLa cells for 24 h was more than 5 mg mL^{-1} [295].

Whereas the reported concentration necessary for cell imaging studies was more than a hundred times less [42].

In an attempt to increase the fluorescence intensity and have negligible cytotoxicity, Zhang *et al.* [48] and Xu *et al.* [283] hydrothermally synthesized CD using lactose as a carbon source and 3-hydroxymethyl amino methane as carbon surface passivating agent. The fluorescence intensity of these CDs was neither affected by the change in the environment nor affected by the salt solution. Hence, they successfully applied these CDs in HeLa cell imaging.

5.2.1.2 Human Breast Cancer MCF-7 Cells and Human Breast Tumor Cells MDA-MB-468

The PPEI-EI passivated CDs were used to obtain two-photon PL images of human breast cancer MCF-7 cells. Cao *et al.* [24] fabricated CD by surface passivating it with PPEI-EI by laser ablation, which had increased water solubility. When this conjugate was used for *in-vitro* bioimaging of human breast cancer MCF-7 cells (incubated in an aqueous buffer at 37 °C) under two-photon luminescence microscopy with excitation by 800 nm laser pulses, MCF-7 cells became bright green upon illumination. The presence of CDs could be seen on the cell membrane and in the cytoplasm of MCF-7 cells. It was found that the CD is translocated through endocytosis and, as expected, translocation of CDs into the cytoplasm depended on the size, surface charge, surface chemistry and temperature (no translocation occurred at 4 °C). Water-soluble CDs prepared from glycine were also used to label MCF-7 and MCF-10A cells. They exhibited different colors of PL images when excited separately at wavelengths of 460–480 nm and of 360–380 nm [296], i.e., blue, green, and red upon excitation at wavelengths of 360–380, 460–480, and 510–530 nm, respectively (Figure 5.5).

Human breast cancer cell MDA-MB-468 has been suggested to be used for cellular uptake by Puvadda *et al.* [297]. The CDs prepared by microwave-assisted method using

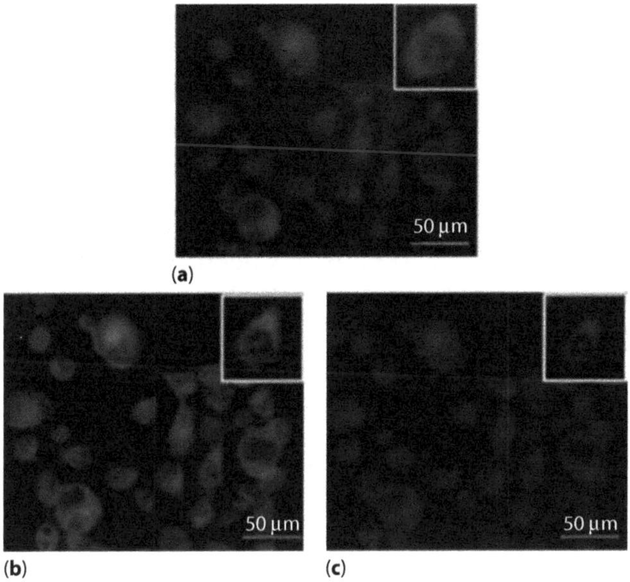

Figure 5.5 PL images of MCF-10A cells treated with CDs (0.5x, 0.18 mg mL^{-1}) under (b) blue (360–380 nm), (c) green (460–480 nm), and (d) red (510–530 nm) light excitation. Insets to (b), (c), and (d): amplified PL images of single cells. (Reproduced with permission from [296]; Copyright © 2013 Royal Society of Chemistry)

aqueous solution of Dextrin in the presence of sulfuric acid exhibited multi-colored luminescent in the entire range of visible light. These water-soluble CDs had the presence of hydrophilic groups on the surface. The CDs were assessed to be biocompatible with MDA-MB-468 cells, and that is why it was suggested that they should be used for cellular uptake and imaging.

5.2.1.3 B16F11 and HEK293 Cells

In cell biology, B16F11 are melanoma cells and HEK293 are human embryonic kidney cells. Both these cells have been [298] labeled with PEG diamine capped CDs and bioimaged under confocal laser scanning microscope. CDs were conjugated with the hyaluronic acid (HA) derivatives by amidation reaction between amino groups on the surface of CDs and carboxylic groups of HA. These HA/CDs were successfully employed for real-time bioimaging of targeted delivery of hyaluronic acid (HA) derivatives to liver.

Using similar conjugate, i.e., hyaluronic acid (HA) as a carbon source and polyethylenimine (PEI) as a passivant, Wang *et al.* [299] fabricated HA-CDs (Figure 5.6) that emit strong blue fluorescence under UV light. Attachment of characteristic residues of HA and PEI to CDs has been confirmed by IR and 1H NMR spectra. This conjugate had much lower cytotoxicity than PEI and high serum tolerance; up to 50 times higher transfection efficiency than that of PEI in the presence of 10% serum. BSA protein adsorption, flow cytometry, and confocal microscopy assays also supported their good performance with

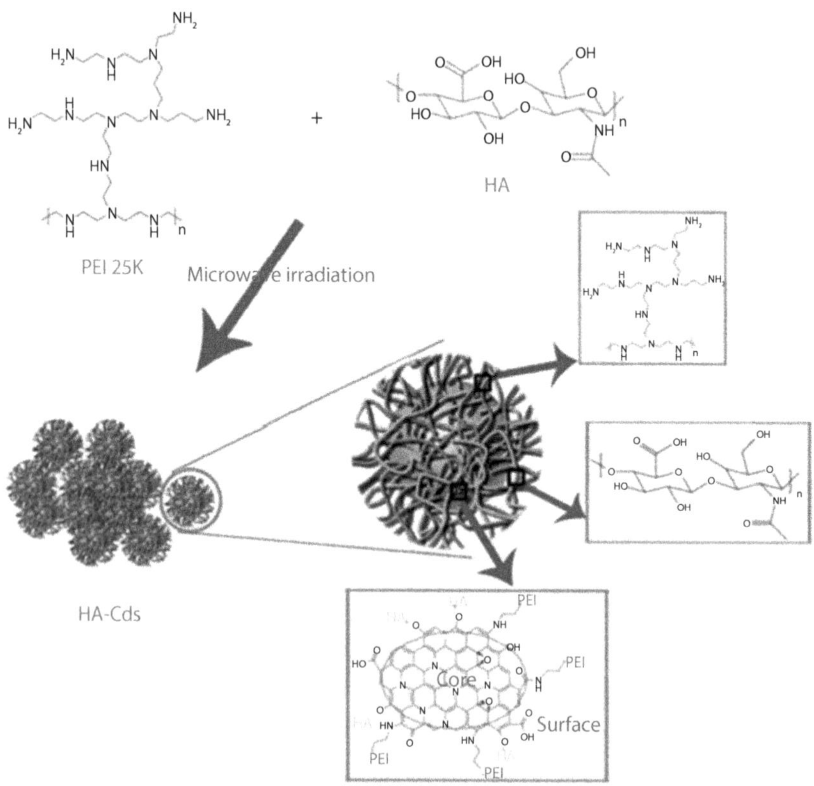

Figure 5.6 Schematic of the synthesis and structure of HA-Cds. (Reproduced from [299]; Copyright © 2017 The Royal Society of Chemistry)

serum and good intracellular imaging ability. Its fluorescence emission is tunable under varying excitation wavelengths. Cancer cells are known to have overexpression of a protein CD44, and HA has an affinity to bind with CD44. Hence, this conjugate may exhibit target cell imaging ability in CD44 over-expressed cells.

5.2.1.4 *Ehrlich Ascites Carcinoma (EAC) Cells*

The synthesis of 2–6 nm size CDs with quantum yield of about ~3% was achieved by oxidation of carbon soot using nitric acid. Use of nitric acid for oxidation induced incorporation of nitrogen and oxygen into CDs, making them water soluble, and exhibited green fluorescence under UV exposure. CDs of various sizes were obtained by this method. Smaller CDs were separated by centrifugation from the mixture of particles of different sizes. Ray *et al.* [42] used this CD for labeling Ehrlich ascites carcinoma cells (EAC) that were collected from the peritoneal cavity of adult female mice after 7 days of inoculation. The labeled cells were separated from free CDs by centrifuging them at 2000 rpm for 3 min. The precipitated cells were suspended in phosphate buffer solution and used for imaging. These CDs could enter into EAC cell cytoplasm without any further functionalization, which helped in tracking their position in the cell using a conventional fluorescence microscope.

5.2.1.5 *Human U87 Cell*

In cell biology, U87 is a human primary glioblastoma *cell* line having epithelial morphology. It was obtained from a stage-four 44-year-old cancer patient. For imaging U87 cells, Wang *et al.* [300] used CD synthesized using vitamin B as precursor via hydrothermal method. This CD had low toxicity and good biocompatibility and fluorescence. The U87 cells were seeded in the medium containing CDs and cultured 2 h. The fluorescence microscopic images exhibited that bright fluorescent CD was distributed only in cell membrane and cytoplasm but not in the nucleus.

5.2.1.6 MGC-803 Human Gastric Cancer Cells

Liu *et al.* [157] conjugated multifunctional ribonuclease A with CD (RNase A@CDs) by a one-step microwave procedure. The fluorescence intensity of RNase A@CDs was QY ~30-fold more than simple CDs. Moreover, RNase A@CDs could penetrate the cell membrane, as well as efficiently enter the nuclei of cells. Interestingly, the photoluminescence intensity gradually decreased as the pH increased with linear dependence, and it reached the maximum at pH 2. This is an important point related to biological systems. A real-time cell electronic sensing (RT-CES) along with MTT assay showed that this conjugate effectively inhibits the growth of human gastric cancer cells (MGC-803). Cell imaging exhibited that it entered the cytoplasm by endocytosis. The RNase A@CDs demonstrated appears to be potential system to image gastric cancer cells *in vivo* and can simultaneously be used for therapy and *in-vivo* fluorescence imaging of nude mice loaded with gastric cancer or other tumors. This work has confirmed that the enzyme played an important role in the penetrability of the CDs.

In another report, CD-based and polyethylenimine (PEI)-adsorbed complexes (CDs@PEI) were applied as cell labeling agent and siRNA nanocarrier for survivin siRNA delivery by Wang *et al.* [100]. The CDs showed excellent water dispersibility, biocompatibility and high quantum yield. In the siRNACdots@PEI, CDs acted as nanocarrier and as fluorescent indicator, while the positively charged PEI acted as the ties attaching negatively charged siRNA to CDs. Confocal fluorescence images revealed the cellular uptake of siRNACdots@PEI complexes and the entrance of siRNA into human gastric cancer cell line (MGC-803), leading to a superior gene silencing efficiency as evaluated by the PCR analysis. The authors suggested that CD-based delivery system may be used to advance the field of siRNA therapeutics and cancer cell labeling.

5.2.1.7 A549 Adenocarcinomic Human Alveolar (Lung) Basal Epithelial Cells

Theranostic requirements, especially for lung-related cells, require highly cyto-compatible, non-cytotoxic material. Hence, Yang *et al.* [301] used A549 human lung adenocarcinoma cells to evaluate the cytocompatibility of the amino-functionalized CDs obtained by hydrothermal carbonization of chitosan and also the viability assessed by MTT (methyl-thiazol-teterazolium) assay (131 MTT assays of cell). The CDs exhibited low cytotoxicity and insignificant toxicity. Hence, the CDs were introduced into A549 cells for their bioimaging using a confocal microscopic test *in vitro*. The photoluminescent spots were observed only in the cell membrane and cytoplasmic area of the cell, but were very weak in the nucleus, indicating that CDs easily penetrated into the cell but did not enter the nuclei.

Bagasse-derived CDs have also been successfully applied in biolabeling of living adenocarcinomic human alveolar basal epithelial cells (A549 cells) by Du *et al.* [302]. No toxicity of bagasse-derived CDs was observed in A549 cells. Fluorescence images revealed that bagasse-derived CDs were located on A549 membrane and in cytoplasm, particularly around cell nucleus, indicating that the CDs might be internalized by endocytosis. Moreover, bagasse-derived CDs were also employed as multichannel fluorescent probes for multicolor tracking agents in flow cytometry.

5.2.1.8 Human Hepatocellular Carcinoma Cells

CDs obtained from glucose by solvothermal method and same CDs surface functionalized with 1,6-hexamethylene diamine are used for labeling of fixed human hepatocellular carcinoma cells by liking as-synthesized CDs with mouse anti-human alpha fetoprotein (AFP) antibody and goat anti-mouse immunoglobulin (IgG), respectively. The indirect labeling approach demonstrated a higher specificity in comparison with direct labeling, and both approaches were found to have strong fluorescence intensity in the labeling of hepatocellular carcinoma cells [55].

Similarly, when Li *et al.* [303] used prepared CD from ginger for human hepatocellular carcinoma cells (HepG2), it provided high inhibition of HepG2 but very low toxicity toward normal mammary epithelial cells (MCF-10A) and normal liver cells (FL83B). Moreover, the inhibition was selective toward HepG2 as compared to other tested cancer cells that were human lung cancer cell line A549, human breast cancer cell line MDA-MB-231 and human cervical cancer cell line HeLa. The CDs upregulated the expression of p53 protein only in HepG2 cell line. An image cytometry exhibited greater uptake of CDs by HepG2 cells, inducing a 18.2-fold increase in intracellular production of ROS when treated with 1.11 mg mL^{-1} CDs in comparison to a negligible increase in ROS of other cells. CDs triggered a pro-apoptotic factor to promote HepG2 cell apoptosis, which inhibited the growth of tumors in nude mice (104 ± 14 vs. 3.7 ± 0.2 mg with and without treatment within 14 days).

5.2.1.9 Kidney Proximal Tubule Cell Line (LLC-PK1)

CDs prepared from waste coffee grounds and incubated at 37 °C for 2 h produces bright PL in the cytoplasmic area and the cell membrane when excited at 800 nm. When used for imaging LLC-PK1 cells, this CD showed it was internalized through endocytosis and got localized in the cell membrane and cytoplasm through endocytosis [304].

Other cancerous cells that have been explored for bioimaging using CDs are: i) *Caco-2 cells*, which are derived from human colon adenocarcinoma (colorectal cancer cells). They have been mostly used for the screening of cytotoxic effects of antitumor drugs and for the study of drug resistance mechanisms; ii) *P19 cells*, which are embryonic carcinoma cell lines derived from an embryo-derived teratocarcinoma in mice; iii) *BGC823 cells*, which are a gastric cancer cell line, etc.

5.2.1.10 C6 Glioma Cells

The C6 glioma cell line is generally used for studying mammalian brain tumors and cell and molecular biology. The C6 cell line, which originated from the glial tumor from the brain tissue of a rat (*Rattus norvegicus*), displays fibroblast morphology as well as glucocorticoid receptors (Figure 5.7). The C6 stem line number is diploid, and the karyotype is stable.

Synthesis of CD is achieved by a simple and facile thermolysis method by pyrolyzing D-glucose and L-aspartic acid, i.e., CD-Asp exhibits tunable full color emission, good biocompatibility and, above all, when injected into the tail of rat, without the aid of any navigating molecule it specifically entered the glioma site of the rat's brain. That means that it can cross the BBB and this CD-Asp could act as a fluorescence imaging agent of brain glioma [305]. The interesting part is that CDs synthesized from D-glucose (CD-G), L-asparic acid (CD-A), or D-glucose and L-glutamic acid (CD-Glu) did not exhibit selectivity for glioma. Therefore, CD-Asp could act as a fluorescence imaging and targeting agent for noninvasive glioma diagnosis (Figure 5.8).

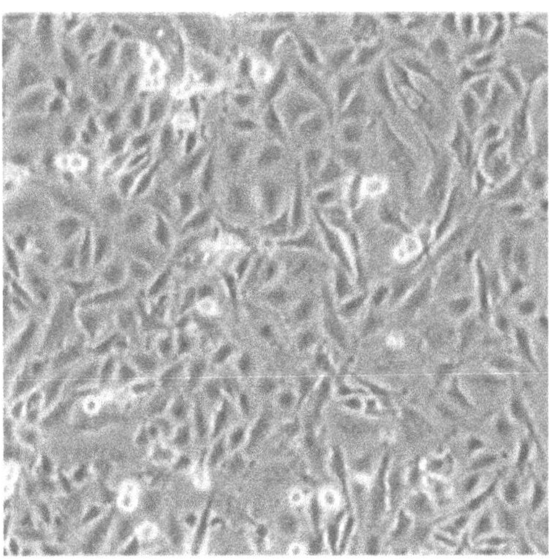

Figure 5.7 C6 glioma cells exhibiting fibroblast morphology.

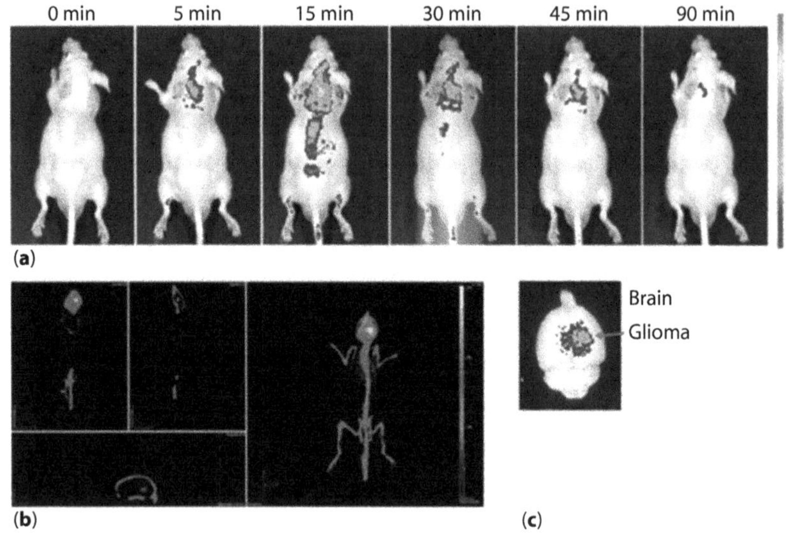

Figure 5.8 *In-vivo* and *ex-vivo* imaging of glioma-bearing mice after tail intravenous injection of CD-Asp. (a) Whole body distribution of CD-Asp as a function of time after injection. (b) Three-dimensional reconstruction of CD-Asp distribution in the brain 20 min after injection. (c) *Ex-vivo* imaging of the brain 90 min after the injection of CD Asp. (Reproduced from [305])

5.2.2 Bioimaging of Nucleus

As we have observed above, CDs get localized in cytoplasm and cell membranes only. CDs cannot easily enter the nucleus. Usually CDs have been found to be mostly localized in the cell membrane, endosomes, lysosomes, mitochondria or endoplasmic reticulum [306]. However, Sharon's group [50] have not only explored the drug carrying and delivery capacity of highly fluorescent sorbitol-derived CDs to HeLa cells, but also did cell imaging which also showed internalization of CDs in the nucleus. They attached folic acid (as a navigational molecule due to its high expression in most cancer cells) and the anticancer drug doxorubicin on the surfaces of the bovine serum albumin-protected CDs (Figure 5.9).

Datta *et al.* [307] have reported that noncovalently attached cationic quaternized CDs to anionic graphene oxide sheet (GO) hybrids act as selective probes with controlled labeling of the

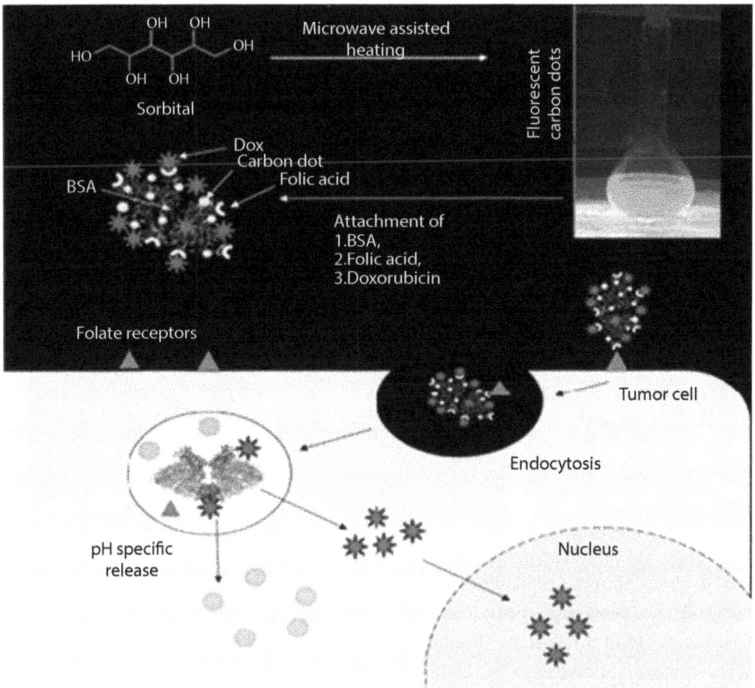

Figure 5.9 Graphic representation of CD-BSA-Dox delivering doxorubicin in cytopkasm and nucleus. (Adapted with permission from [50])

cell nucleus or cytoplasm, depending on the CD loading (Fig. 5.10a,b). GO-CDs hybrid with different mass ratios were evaluated for selective cytoplasm/nucleus labeling. The presence of a larger amount of CD than GO caused high fluorescence. They believed that this was the reason that led to a nearly homogenous coating of GO surface with CDs. They demonstrated the selective cell labeling of CD in mouse fibroblast NIH/3T3 cells incubated for 24 h with pure CDs (Figure 5.10c) and with hybrid GO-CDs conjugate. CDs mostly entered the nucleus rather than cytoplasm, which they suggested was due to the presence of strong positive charge, which originated from the quaternary ammonium precursor employed for CDs synthesis. The CDs were internalized by endocytosis and then cross the nucleolemma, leading to large amounts of CDs in the nuclei

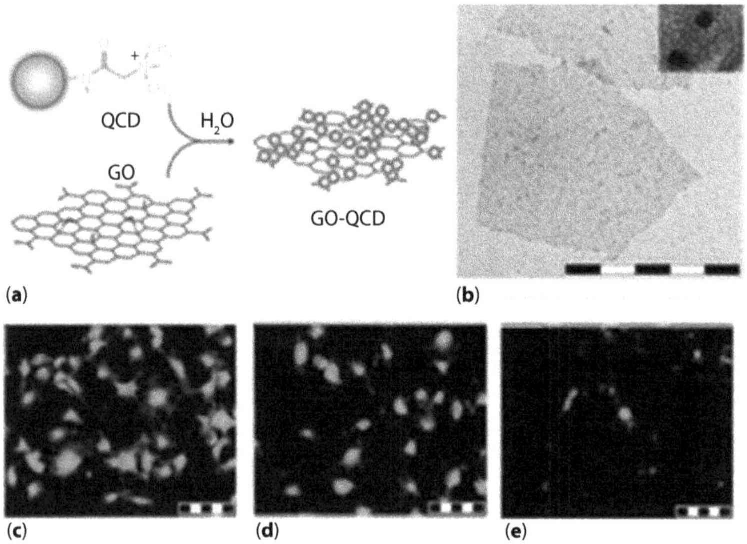

Figure 5.10 (a) Schematic representation of the self-assembly process of graphene oxide-cationic quaternized carbon dots (GOQCD) hybrid materials in water. (b) Transmission electron microscopy image of the hybrid material GO-QCD (with a GO:QCDs ratio of 0.5, coded as GO250QCD500). The scale bar in (b) corresponds to 1 mm, the inset (top right) shows the high-resolution transmission electron microscopy image. Fluorescence images of mouse fibroblast NIH/3T3 cells containing: (c) quaternized carbon dots (QCDs); hybrid materials GO-QCD with a GO:QCDs ratios of (d) 0.5; (e) 1. Concentration used: 100 g/mL, time of incubation with cells: 1 day. Scale bars: 200_m. (Modified with permission from [307]; Copyright © 2014 Royal Society of Chemistry.)

(Figure 5.10c). The cell labeling depended on the ratio of CD and GO in the hybrid (Figure 5.10).

5.2.3 Bioimaging of Virus

More than a decade ago, Stroh *et al.* [149] had suggested that In HIV-1 CDs can overcome cellular membrane barrier more easily by bonding to membrane translocation peptides TAT (HIV-1 trans-activator of transcription). In this way more CDs can accumulate in the cell and offer better efficiency of the intracellular labeling, as well as better imaging.

The effects of CDs on viruses and highly efficient labeling of viruses with QD are still largely unknown. According to Lv *et al.* [308], this is because of the lack of prerequisite for the long-term tracking of virus invasion at the single virus level to reveal mechanisms of virus infection. The viral envelope lipids are hard to be labeled with QDs due to the lack of efficient methods to modify viral envelope lipid and maintain the intactness and infectivity of labeled viruses. Therefore, they developed a method to label viral envelope lipids with QDs by harnessing the biotinylated lipid-self-inserted cellular membrane. Using the pseudorabies virus (PrV) they demonstrated that biotinylated lipids can spontaneously insert in cellular membranes of host cells during culture and then be naturally assembled on the progeny of pseudorabies virus (PrV) via propagation. Moreover, the biotinylated PrV can be labeled with streptavidin-conjugated QDs, with a labeling efficiency of ~90%. Such lipids labeled with QDs retain the intactness and infectivity of labeled viruses to the largest extent. This will help in studying the mechanisms of virus infection at the single virus level.

Du *et al.* [309] also investigated the effect of CDs on the replication of pseudorabies virus (PRV) and porcine reproductive and respiratory syndrome virus (PRRSV) as the models of DNA virus and RNA virus, respectively. After the analyses of virus titers and the expression of viral proteins, it was found that cells treated with CDs can significantly inhibit the multiplication of PRV and PRRSV by inducing interferon-α (IFN-α) production and the expression of IFN-stimulating genes (ISGs), which in turn inhibits virus replication.

5.2.4 Bioimaging of Bacteria

Prior to the use of semiconductor QD [310], bacterial labeling was often done by fluorescent dyes. The first attempt of imaging and labeling of *Escherichia coli* cells by CD was done by Sun *et al.* [56]. They used $PEG_{1500}N$-passivated and PPEI-EI-passivated CDs. This development of a new approach for simple, efficient, and inexpensive detection of a broad range of bacterial species is an advanced technology where CDs can be used for bacteria imaging, labeling and detection, which now has several applications in healthcare and diagnostics [77, 74].

Another approach by Sachadev *et al.* [166] was using chitosan as carbon source, where CDs were synthesized by pyrolyzing it in a microwave and then passivating it with PEG. These CDs are 5–8 nm in diameter. When used for labeling recombinant green fluorescent protein (GFP)-expressing *E. coli*, CDs showed blue, light green and red color fluorescence, whereas GFP *E. coli* exhibited bright green under blue light excitation and were fairly visible under UV light excitation. CDs-labeled GFP *E. coli* showed blue, light green and red color emissions. Based on the results of the GFP-*E. coli* model the authors suggested the optimal labeling time and the mechanism involved, and attributed it to the quenching of the inherent fluorescence of green fluorescent protein in GFP-*E. coli* by CDs.

In a similar approach, *E. coli* was labeled using CDs that were surface functionalized with hydrocarbon chains. These amphiphilic CDs have high affinity to bacterial cell membranes. For attachment of CDs, *E. coli* was incubated with hydrocarbon chain functionalized CDs that yielded strong fluorescence. The authors [161] suggested that it might be used for selective detection of bacteria in cases of mixed populations of bacteria. As can be seen in Figure 5.11b showing the fluorescence microscopy images of *E. coli* incubated with amphiphilic CDs, the broad range of excitation and emission of CDs reflects the multicolor fluorescence.

Weng *et al.* [311] used mannose-modified fluorescent CDs obtained from mannose and ammonium citrate for labeling *E. coli*. These CDs were water soluble, having average size of 3.1 nm and QY of 9.8 at excitation wavelength at 365 nm and emission at 450 nm. The mannose-CDs could link to the FimH lectin unit in the flagella of wild-type 1 *E. Coli* K12 strain, allowing quantitative bacteria detection. The detection level was 450 colony-forming unit per mL. The efficacy of mannose-CD to label *E. coli* was tested in tap water, human urine, and juice.

Pseudomonas aeruginosa and *Fusarium avenaceum* fungi were probed for bioimaging using fluorescent CDs synthesized from *Punica granatum* fruit by hydrothermal method.

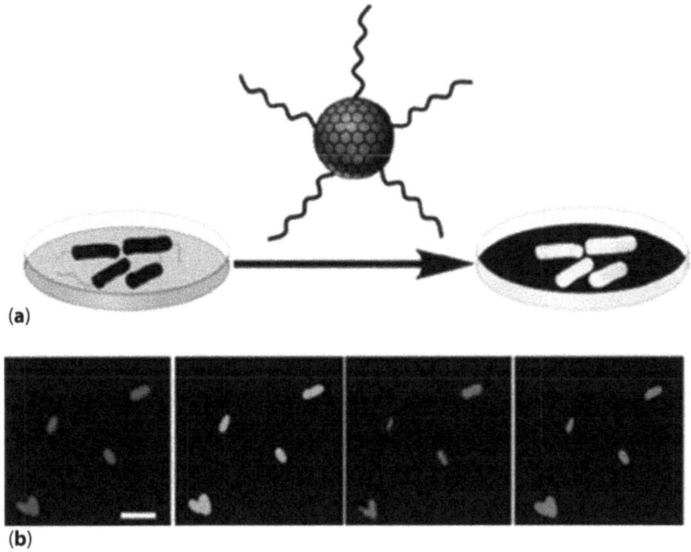

Figure 5.11 Amphiphilic CDs for bacteria labeling. (a) Schematic representation of the labeling methodology of bacteria with fluorescent CDs; (b) Multicolor fluorescence microscopy images of *E. coli* incubated with amphiphilic CDs at different excitation/emission wavelengths. Blue: excitation at 365 nm, emission filter 420 LP; green: excitation at 470 nm, emission filter 510 LP; magenta: excitation at 510 nm, emission filter 545 nm; red: excitation at 540 nm, emission filter 605 nm. Scale bar corresponds to 5_m. Exposure time was 0.5 s in all experiments. (Reproduced with permission from [161]; Copyright © 2015 Royal Society of Chemistry)

The CDs showed a strong emission at 453 nm when excited at 383 nm. These CDs were tested for toxicity using *Bacillus subtilis.* It was found that they were neither toxic nor inhibited the growth of *B. subtilis* [284]. When incubated with both the bacteria and fungi the CDs were rapidly internalized and exhibited two color (green and red) images under confocal fluorescence microscope.

5.2.5 Bioimaging of *Drosophila melanogaster*

Ghosh *et al.* [312] prepared fluorescent CDs by pyrolyzing wood wool at 600 °C in a muffle furnace in the presence of nitrogen/oxygen gas flow and then purified it. The structure that they obtained was nano-onion having fluorescent CD-like properties. They administered it orally to *D. melanogaster* and did the imaging of the full life cycle of larval organisms, including *D. melanogaster,* without any toxic effect. They claimed that owing to its interaction with the biomolecules, the fluorescence of this material was enhanced *in vivo.*

5.3 CDs as Sensor

For CDs to be used as sensor, it is of particular interest and significance that CDs can exhibit PL emission in the near-infrared (NIR) spectral region under NIR light excitation. It should be noted that NIR PL emission of CDs excited by NIR excitation is particularly significant and useful as a sensor or *in-vivo* bionanotechnology because of the transparency of body tissues in the NIR "water window." Interestingly, the PL from CDs can be quenched efficiently by either electron acceptor or electron donor molecules in solution, indicating that photoexcited CDs are excellent electron donors and electron acceptors. Mechanistically, the fluorescence emissions in CDs are attributed to radiative recombination of the carbon particle surface-trapped electrons and holes, where the large surface (relative to the particle volume) and diverse surface energy trapping sites in the small carbon nanoparticles are stabilized by the surface passivation agents. An important property of CD is that, unlike other semiconductor QDs containing heavy metals such as cadmium, carbon is generally not considered as a toxic element.

The fundamental principle of biosensing using CD is that assaying PL quenching makes the CD-based sensor/detector very sensitive and selective. Here, quenching refers to decreases in the fluorescence intensity of CD due to FRET. Because CDs are low in cytotoxicity they have good stability and cell internalization, high sensitivity, specificity, and accuracy; and are versatile material for designing multifunctional biosensors to detect biological pH, metal ions, proteins, sugars, vitamins, DNA, etc.

5.3.1 Intracellular Detection of Ions

Because of excellent electron donor or electron acceptor capability, CDs are used in sensor for detection of metals in ionic form and their concentration. A number of CDs synthesized from different precursors have been developed as direct fluorescent sensors for metal ions. Fluorescence "turn-off" assays have been used for the detection of various metal ions, including Hg^{2+}, Cu^{2+}, Pb^{2+}, Fe3^{+}, Ag^{+} and Cr^{6+} [313, 314]. Synthesis of CD using different precursors has been used for detection of various ions such as to detect $Cu^{2+,}$ $Hg^{2+,}$ Ag^{2+}, Cr^{3+}, $Fe^{3+,}$ K^{+}, Cl^{-}, and H^{+}. The mechanisms of ion detection by CD is the same for all ions, i.e., the functional groups on the surface of CDs act as an indicator that has distinctive affinities to different ions, which quenches the PL intensity of CD through an electron or energy transfer process. Overall, the detection mechanism of heavy metals by CDs biosensors is based on the interaction between the functional groups on the CDs surface with the target ions. This interaction results in the quenching of the fluorescence of CDs by metal ions through an energy or electron transfer process. Therefore, CDs can be designed to detect several metal ions.

5.3.1.1 Detection of Ag Ions

Li *et al.* [117] obtained CDs from carbon soot from a candle that were used as a sensing platform to detect silver (Ag^{+}) ions with a limit of detection as low as 500 pmol/L. Later, water-soluble CDs synthesized by one-step heat treatment of ethylene glycol solution, which involved hydration, crosslinking and carbonization processes, exhibited green luminescence under UV excitation and were used for the detection of Ag^{+} ions. The interesting part of this sensing method was that instead of quenching the luminescence of CD, Ag^{+} exhibited an enhancement effect on the photoluminescence of CDs. Gao *et al.* [315] attributed it to the reduction of Ag^{+} to silver nanoclusters (Ag^{0}) on the surface of the CDs. There was a linear relationship between fluorescence intensity and concentration of Ag^+ ions. This CD can be used for sensitive and selective detection of silver ions in aqueous environment with a detection limit of 320 nM and a linear range of 0–90 μM.

Nitrogen-doped CDs synthesized by a one-pot solvothermal method using different nitrogen contents exhibited increased quantum yield and the highest emission efficiency of up to 36.3%. The fluorescence enhancement may have been due more to polyaromatic structures induced by incorporating nitrogen atoms and protonation of nitrogen atoms on dots. This N-doped CD can be used as a multifunctional fluorescence-sensing platform because apart from Ag they can also be used to detect pH values and Fe^{III} ions in aqueous solution. Its fluorescence intensity is inversely proportional to pH values across a broad range from 5.0 to 13.5. The quantitative determination of Ag^I can be accomplished with N-doped CDs by using the linear relationship between fluorescence intensity of N-doped CDs and concentration of Ag^I. The doping of nitrogen into CDs can not only greatly improve their fluorescence performance, but also acts as an active site to selectively and sensitively detect pH values, Ag^I, Fe^{III}, and H_2O_2 in aqueous solution (Figure 5.12).

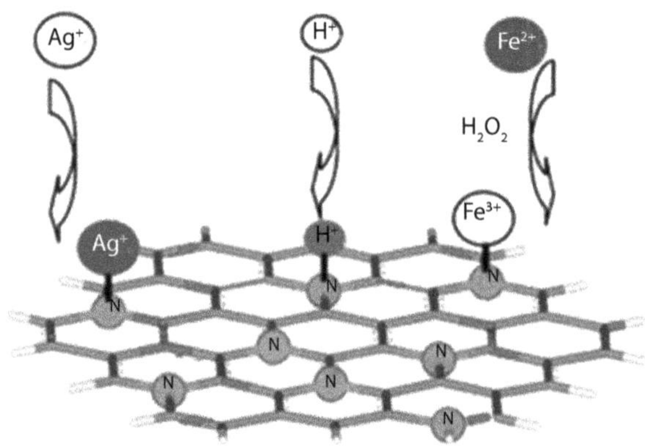

Figure 5.12 Nitrogen-doped CDs with high fluorescence as an active site, selectively and sensitively detects pH values, Ag^I, Fe^{III}, and H_2O_2. (Adapted from [315])

5.3.1.2 *CD for Detection of Cu Ion*

A selective Cu^{2+} sensor based on the CDs functionalized with branched poly(ethylenimine) (BPEI) that could quench the fluorescence signal was developed by Dong *et al.* [316]. As represented in Figure 5.13, the amino groups on the surface of functionalized CDs can capture Cu^{2+} ions leading to cupric amine and thus selectively quenching florescence of CDs via an inner-filter effect (Figure 5.13). The low detection limit of this sensor is 6 nmol/L, with a dynamic range of 10 to1100 nmol/L.

Several attempts have been made to prepare CDs for intracellular sensing of Cu^{2+} ions. Zhu *et al.* [317] successfully integrated CDs with the organic molecule *N*-(2-aminoethyl)-*N*,*N*,*N*'tris(pyridin-2-ylmethyl)ethane-1,2-diamine (AETPEA) and CdSe/ZnS QD for intracellular sensing Cu^{2+} ions. The obtained fluorescent CDs were able to monitor Cu^{2+} ions in the range of 5–200 μmol/L under physiological conditions. CDs were able to detect cellular Cu^{2+} ions and real-time imaging in living cells was exhibited by the presence of CDs in various organelles of the cell compartments. The exogenous administration of Cu^{2+} changed the fluorescence emission color of CDs from green yellow to red.

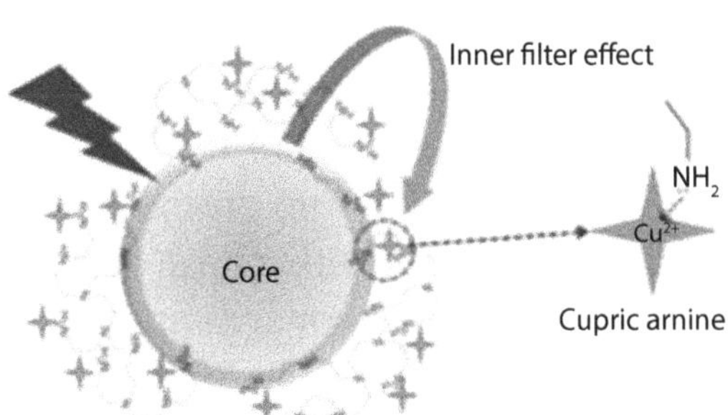

Figure 5.13 Schematic representation of the fluorescence of the branched poly(ethylenimine) (BPEI)-functionalized CDs (BPEI-CDs) quenched by Cu^{2+} ions. (Reproduced with permission from [316]; Copyright © 2012 The American Chemical Society)

A selective and sensitive optical sensor for Cu^{2+} using CD was fabricated by Salinos-Castillo *et al.* [116]. The CD was prepared by microwave-assisted method using citric acid in the presence of polyethylenimine (PEI). Dynamic light scattering (DLS) data exhibited well-dispersed 12 nm (0.212 PDI) size water-soluble CD showing typical bands of PEI in the IR spectrum; the zeta potential at pH 4 was +21 mV and the ninhydrin colorimetric assay confirmed the presence of primary and secondary amines on the surface of CDs. The CDs have blue luminescence under 265 nm UV excitation and fluorescence quantum yield of 30%. These CDs exhibit good upconversion fluorescent properties along with strong emission luminescence in the 450–650 nm range. All these observations of CD were useful in applying it for intracellular sensing as well as imaging of Cu^{2+} in biological system. It had maximum fluorescence activity at pH 4, which is attributed to the protonation of amino groups. The quenching behavior of Cu^{2+} ions in the luminescence properties of the CD is due to the inner-filter effect. CDs having down- and upconversion luminescence properties are suitable for selective and sensitive detection of Cu^{2}. In addition, these fluorescent nanosensors show low cytotoxicity and good cell permeability.

Using a similar strategy, Liu *et al.* [318] synthesized a ratiometric nanosensor based on dual-emission CDs for Cu^{2+} ions detection. CDs were synthesized from *N*-(β-aminoethyl)-γ-aminopropyl methyldimethoxysilane (AEAPMS) and their surface was coated with Rhodamine B-doped silica nanoparticles (Figure 5.14). This strategy resulted in the presence of residual methoxysilane and ethylenediamine groups on the surface of CDs, allowing Cu^{2+} recognition with a limit of detection of 35.2 nmol/L. The CD-based nanosensor was applied for the ratiometric fluorescence imaging of Cu^{2+} in MCF-7 cells and also for the real-time quantification of Cu^{2+} in tap water.

Another Cu^{2+} sensor was developed by Zong *et al.* [319]. They selected spherical mesoporous silica as nanoreactor added to the solution of citric acid and three other inorganic salts (NaCl,

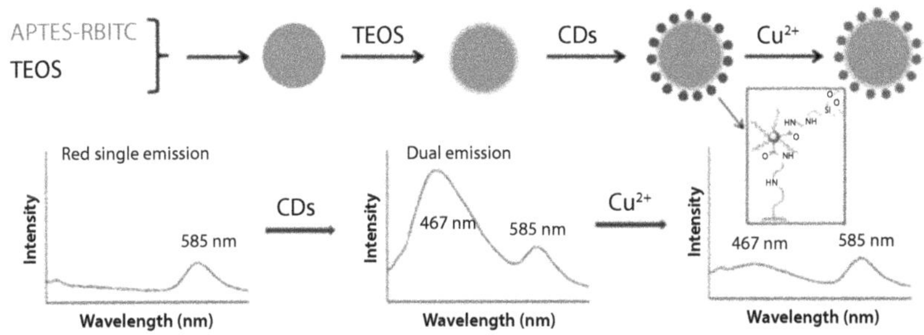

Figure 5.14 Schematic representation of the ratiometric fluorescent Cu2+ nanosensor based on CDs@Rhodamine B-doped silica nanoparticles. (Reproduced with permission from [318]; Copyright © 2014 The American Chemical Society)

LiCl, and KNO_3), and ultrasonificated it to obtain CDs. The CDs bonded specifically with Cu^{2+} and quenched the fluorescence of CDs. When L-cysteine was added to the solution of CDs-Cu^{2+}, Cu^{2+} was released from the surface of the CDs, thereby recovering the fluorescence of CDs. This sensor can detect both Cu^{2+} and L-cysteine; the detection limit for Cu^{2+} was 2.3 × 10–8 mol/L and for L-cysteine was 3.4 × 10–10 mol/L.

Another CD-based probe for Cu^{2+} ion was fabricated using CD synthesized from prawn shell. These CDs were 4 nm in size, monodispersed, emitted blue light under UV light, were water soluble, had good stability and high quantum yield of 9%. When used for sensing Cu^{2+} ions in sea water, drinking water and river water it exhibited excellent selectivity and detection limit of 5 nM [320]. As we know, pH affects the sensing technique and PL intensity, and quenching the PL intensity of CDs; therefore, the appropriate pH to get the best required CDs was tested. It was found that pH from 2 to 4 showed maximum PL intensity; therefore, it is accepted as the best range for sensing. The addition of copper ions leads to the quenching of CDs PL intensity over the full range of pH from 2 to 12. The optimum (higher PL quenching) pH was 4. Beyond this value, the PL quenching efficiency was relatively lower, which may be due to the protonation of CDs' surface functional groups such as $-NH_2$. The authors have suggested that these protonated functional groups are arduous in forming complexes with the copper ions due to the electrostatic repulsion between the protonated CD surface functional groups and the positive charges of copper ions (Cu^{2+}). Whereas in alkaline solution (pH > 7), the copper ions do not influence the CDs, hence show less quenching. The reaction time required for the completion of the reaction between CD and Cu^{2+} ions is 10–15 minutes. The selectivity of Cu^{2+} was tested against various metal ions like Zn^{2+}, Ni^{2+}, Hg^{2+}, Cu^{2+}, Fe^{2+}, Ni^{2+}, Pb^{2+}, Ca^{2+}, Mn^{2+}, and Cd^{2+} and it is observed that the Cu^{2+} caused the highest quenching, whereas not much quenching was seen in the presence of other metal ions.

5.3.1.3 Detection of Fe^{3+} Ions

Ferric ion, Fe^{3+}, is also fluorescently detected by CDs. Dopamine is supposed to have a special affinity for Fe^{3+}. Therefore, Qu *et al.* [321] have used dopamine as the raw material for synthesizing CDs to be used as a Fe^{3+} sensor with a detection limit of 0.32 μmol/L.

Another approach of detecting Fe^{3+} ion was to generate blue luminescent CDs with a high quantum yield of 31% using a solid phase synthesis. Zhang *et al.* [322] suggested that the strong coordination of Fe^{3+} ion to oxygen-rich groups on CDs' surface induces fluorescence quenching by nonradiative electron transfer (Figure 5.15), thus offering a fluorescence "turn-off" avenue for the determination of Fe^{3+} ion.

The above-mentioned approach of Zhang *et al.* [322] is based on the well-known fact that nitrogen (N) is a popular atom for synthesis of heteroatom-doped CDs due to the similarity of atomic size of N and C and the strong bonding between them. The N doping of CDs not only enable enhanced photoluminescence (PL) efficiency and tunability of PL emission, but also enrich surface functional groups to broaden its application. Therefore, NDs are being tried in various applications. Use of

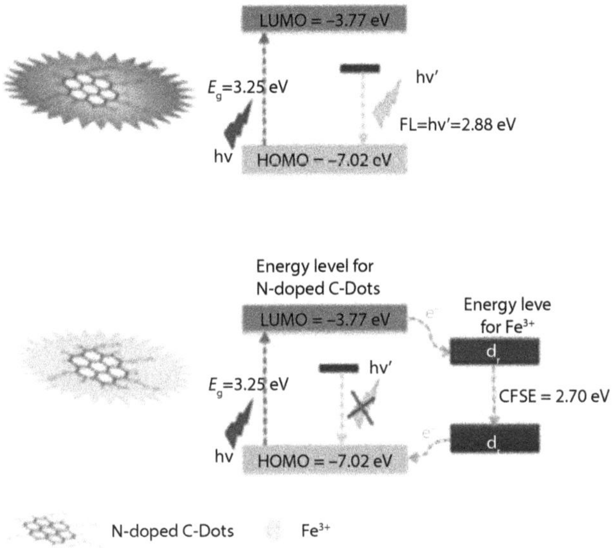

Figure 5.15 Sensing principle of the N-doped CDs-based probe for Fe^{3+}. (Adapted with permission from [322])

N-doped CD for detecting Fe^{3+} in cell, as well as for imaging, is also possible by synthesizing NCD from the antibiotic drug amino salicylic acid. The as-prepared NCDs possess tunable PL and have a quantum yield of 16.4%. The cellular toxicity on H1299 cancer cells has been tested and found to be nontoxic. Moreover, this drug-derived NCD exhibits excellent sensitivity in detection of Fe^{3+} in living cells.

5.3.1.4 *Detection of Hg^{2+} Ions*

Since Hg^{2+} is one of the most toxic heavy metal ions in the environment, it has received attention for developing CD-based methods to detect Hg^{2+}. Direct use of CDs as fluorescent sensors for label-free detection of Hg^{2+} and bithiols through an effective electron transfer process has been reported based on the fact that Hg^{2+} ions are able to quench the fluorescence of QD.

Like Cu^{2+}, Hg^{2+} ions can also be selectively detected by CDs synthesized by laser ablation and functionalized with

NH_2-polyethylene-glycol (PEG_{200}) and *N*-acetyl-*L*-cysteine (NAC). These functionalized CDs are fluorescent with excitation at 340 nm and emission at 450 nm. Hg^{2+} ions were found to be detected by the functionalized CDs via fluorescence quenching [58].

A new type of CD from grapefruit peel was prepared through hydrothermal method that can effectively quench the fluorescence of CDs, having Hg^{2+} detection limit of 0.23 nM. Lu *et al.* [323] have applied it to detect Hg^{2+} in the river. Later, Liu *et al.* [324] also detected Hg^{2+} ion in the river, lake and tap water. The CDs they prepared had excellent fluorescent stability and a very high detection limit of 1 fM. The CD was synthesized using polyethylene glycol (PEG) refluxed with NaOH. To bind Hg^{2+} ions specifically to CDs, Zhang and Chen [325] used vitamin B as carbon and nitrogen sources to prepare a highly fluorescent, nitrogen-rich CD by hydrothermal method. The successful use of this fluorescent probe established that Hg^{2+} can be detected up to a detection limit of 0.23 nM. So, we have seen that the interaction between CDs and Hg^{2+} or other ions results in the quenching of PL intensity. The PL is recovered when the interaction is broken via external force. Recently, a fluorescent turn-on system was also designed. Cui *et al.* [267] fabricated novel oligodeoxyribonucleotide (ODN)-CDs and graphene oxide optical sensor to detect Hg^{2+} ion. The fluorescence of ODN-CDs was efficiently quenched by graphene oxide via fluorescence resonance energy transfer (FRET). With the addition of Hg^{2+}, the luminescence was recovered by releasing ODN-CDs from GO, due to the formation of T-Hg^{2+}-T duplex.

5.3.2 Detection of Small Molecules

During the development of metal ion sensors, scientists have also expanded the use of CDs as probes for detection of small molecules. Unlike heavy metal ions detected by fluorescence quenching mechanism CDs, small molecules are generally detected by restoring fluorescence of the quenched CDs. It is also used for some metal ions; for example, Du *et al.* [302] successfully detected the iodide in aqueous solution by replacing the Hg^{2+} ions bound with CDs.

5.3.2.1 Detection of Nitric Oxide

In biological system, reactive oxygen species and their precursor can be selectively detected by surface modified CDs. The free radical nitric oxide is an important biomolecule with several physiological and pathophysiological roles [326–329]. Therefore, Yu *et al.* [330] developed a highly selective CD-based nanosensor to detect nitric oxide in living cells and in water. The mechanism of nitric oxide detection is based on the insertion of the nitric oxide probe, phenylenediamine containing naphthalimide, on CDs. In the presence of nitric oxide, the CD-based sensor would accept energy, leading to a fluorescence change from blue to green.

5.3.2.2 Detection of Phosphate

In another report by Zhao *et al.* [199], phosphate (Pi) was detected by CD-based sensor with the aid of Eu^{3+} ions (Figure 5.16), which has high affinity with carboxylate groups of Pi molecule. The sensing mechanism is based on the initial decrease of the fluorescence of CDs due to the coordination of Eu^{3+} ions with the carboxylate groups of CDs, leading

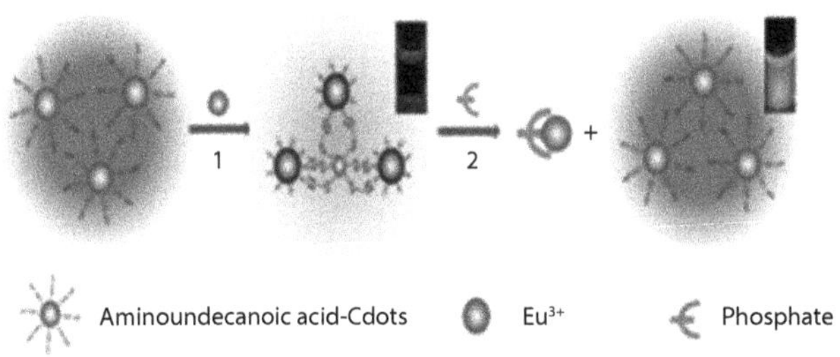

Figure 5.16 Diagram for sensing of phosphate based on the modified CDs mediated by Eu^{3+} ion. (Reproduced with permission from [199]; Copyright © 2011 Royal Society of Chemistry)

to the CDs' aggregation. The dissociation of CD aggregates would occur upon the addition of Pi, recovering the fluorescence, which is used for the detection of Pi. An ultrasensitive, highly blue fluorescent CDs-based sensor was achieved via a one-step hydrothermal treatment of potatoes to detect phosphate [283]. This CD was highly biocompatible and had very low cytotoxicity. This CD-based probe, apart from detecting phosphate, has also been applied for bioimaging of HeLa cells.

5.3.2.3 Detection of Reactive Oxygen Species

In a similar approach as for nitric oxide, for detection of reactive oxygen species, CDs were functionalized with a specific superoxide ($O_2^{\bullet -}$) target, hydroethidine (HE), to get a hybrid CD-HE detecting $O_2^{\bullet -}$. The ratiometric detection of $O_2^{\bullet -}$ occurs upon reaction of CD-HE with $O_2^{\bullet -}$, leading to an emission peak due to the reaction products of HE with $O_2^{\bullet -}$ at 610 nm. This CD-based $O_2^{\bullet -}$ fluorescent biosensor was highly sensitive, having a low detection limit of 100 nmol/L, dynamic linear range from 5×10^{-7} to 1.4×10^{-4} mol/L and high selectivity for $O_2^{\bullet -}$ over other reactive oxygen species and metal ions [331].

5.3.2.4 Detection of H_2S

For detection of H_2S a naphthalimide azide derivative anchored CD has been synthesized. Using this CD a FRET ratiometric fluorescent sensor was developed by Yu's group [330], which could detect H_2S with a detection limit of 10 nM, which is the lowest among the fluorescent H_2S sensors. This highly sensitive sensor could be employed to detect H_2S not only in aqueous media and serum, but even inside living cells, which further broadened the applications of CDs in the biomedical area. In this sensor, CDs serve as the energy donor and also the anchoring site for the probe (Figure 5.17). This sensor is highly selective.

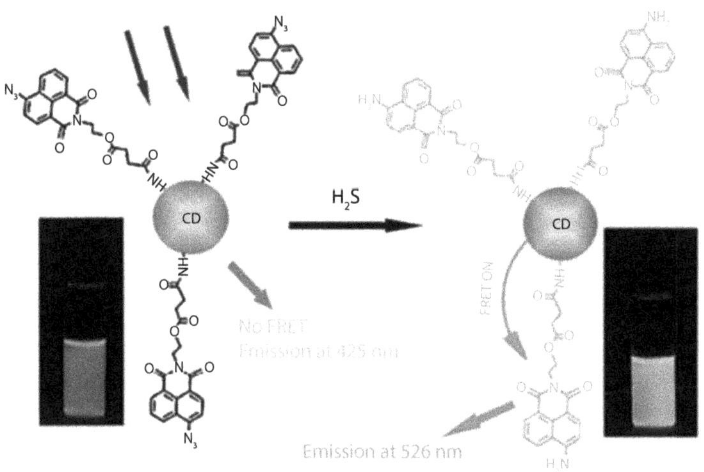

Figure 5.17 Schematic representations of the sensing of H_2S (Reprinted with permission from [330]; Copyright © 2013 Royal Society of Chemistry)

5.3.2.5 *Detection of TNT*

Trinitrotoluene (TNT) $C_6H_2(NO_2)3CH_3$ is known as an explosive material.

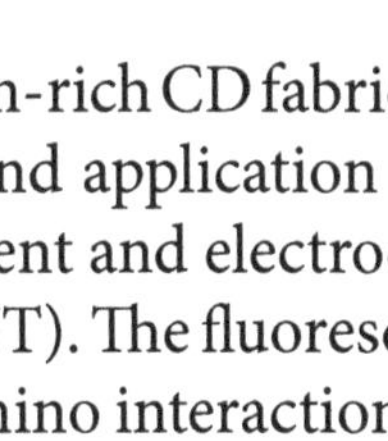

Chemical formula of TNT $C_6H_2(NO_2)_3CH_3$

A nitrogen-rich CD fabricated via a microwave-assisted pyrolysis has found application as a dual sensing platform for both the fluorescent and electrochemical detection of 2,4,6-trinitrotoluene (TNT). The fluorescence sensor was developed based on the TNT-amino interaction, which quenched the luminescence of amino-functionalized CDs via charge transfer [332].

5.3.2.6 *Detection of Hydroquinone*

Hydroquinone (Benzene-1,4-diol or quinol), is an aromatic organic compound that is a type of phenol $C_6H_4(OH)_2$. Ni *et al.* [333] were the first to develop a CD-based fluorescent

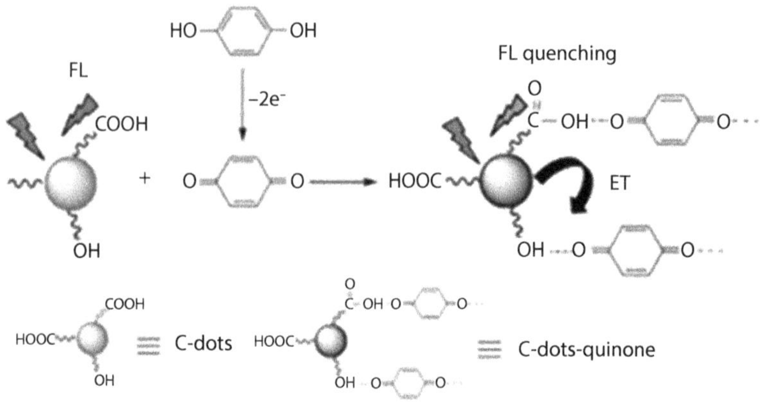

Figure 5.18 Graphical abstract of a facile method based on CDs for sensitive detection of H_2Q as proposed by Ni *et al.* (Reproduced with permission from [333]; Copyright © 2015 Elsevier)

sensor to detect hydroquinone in waste water without using any enzyme (Figure 5.18).

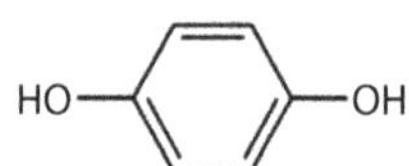

Chemical formula of hydroquinone

Hydroquinone can directly quench the fluorescence of CD. They proposed that the quenching was due to electron transfer from CDs to oxidized hydroquinone. The detection limit of this CD-based probe was down to 0.1 μM. Moreover, this method showed high selectivity for hydroquinone as compared to many potential interfering species.

5.3.2.7 *Detection of Surfactant*

Surfactants are compounds that lower the surface tension between two liquids, between a gas and a liquid, or between a liquid and a solid such as detergents, wetting agents, emulsifiers, foaming agents, and dispersants. Utilizing the substantial fluorescence of CDs, Bhaisare *et al.* [334] employed it as a fluorophotometric sensor to determinate the critical micelle concentration of ionic and nonionic surfactants via Stokes shift

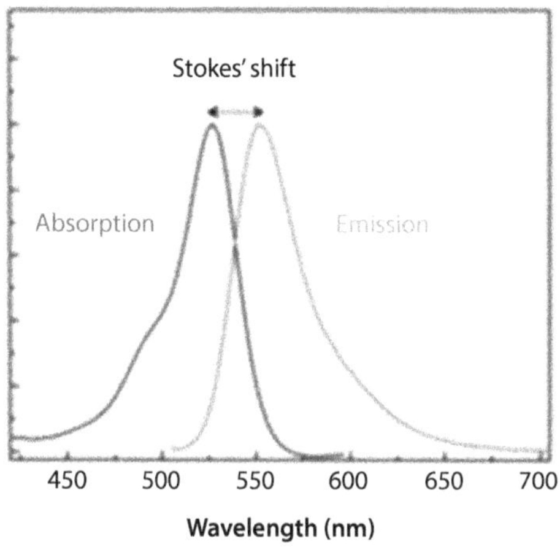

Figure 5.19 Stokes effect. (Source: Wikipedia)

(Figure 5.19). Stokes shift is the difference in wavelength or frequency units between positions of the band maxima of the absorption and emission spectra, e.g., fluorescence and Raman spectra of the same electronic transition.

5.3.2.8 Detection of Humidity

A humidity sensing device has been fabricated using CD film. CD was obtained by electrochemical ablation of graphite. The conductivity of the CD film shows a linear and rapid response to relative humidity. The oxygen-containing groups on the surface are essential for the humidity sensing property of CDs. This CD prepared by Zhang *et al.* [335] possesses abundant oxygen-containing groups (carboxyl and hydroxyl groups) on the surface. They have suggested that when the humidity of the ambient environment is very low, the electrons migration between neighboring CDs depends on the jumping mode. Only electrons with very high energy can do the leap migration (Boltzmann's law). Hence, the conductivity of the solid CD films in dry environment is very low. Whereas, in humid environment the oxygen-containing groups on the surface of the neighboring CDs form hydrogen bonds with water molecules,

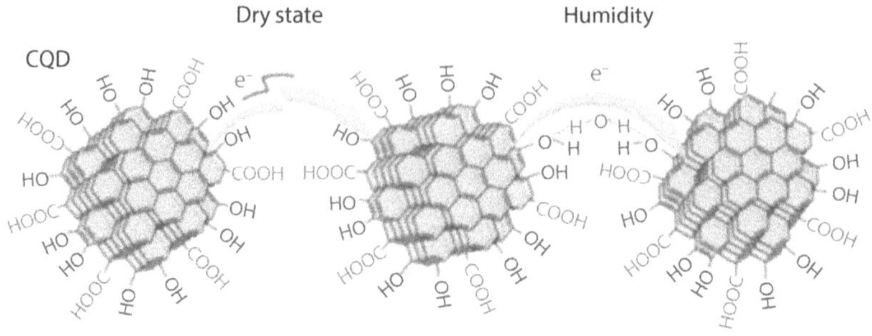

Figure 5.20 Schematic diagram of the electrons' migration routes. (Adapted with permission from [335])

which promote the electrons migration, thus increasing the conductivity. Therefore, the mechanism of humidity sensor is the formation of hydrogen bonds between CDs and water molecules in the humidity environment that promote the electrons migration (Figure 5.20).

5.3.3 Detection of Biological pH Value

The physiological pH value in living cells and tissues is also determined by CDs because the fluorescence intensity of the CD is highly sensitive to pH. Li *et al.* [37] have shown that the CDs exhibit strong emission at pH 7.0; whereas, under strongly alkaline or acidic conditions, the PL is nearly completely quenched. The reason for this is that there are carboxylic acid groups on the surface of CDs and due to the ionization of carboxylic-acid groups, the local surface environment is affected. This was proved when they synthesized CD using graphite rods as both anode and cathode, and NaOH/EtOH as electrolyte, by a one-step sodium hydroxide-assisted electrochemical treatment with a current intensity of 10–200 mA cm^{-2} (Figure 5.21). These CDs were highly fluorescent, emitted bright and colorful photoluminescence covering the entire visible spectral range and had excellent upconversion fluorescent, stable ionic, long fluorescence lifetime and pH-sensitive photoluminescence properties. As a reference, they also used sulphuric acid instead of NaOH, e.g., H_2SO_4/EtOH, as electrolyte; but this did not yield

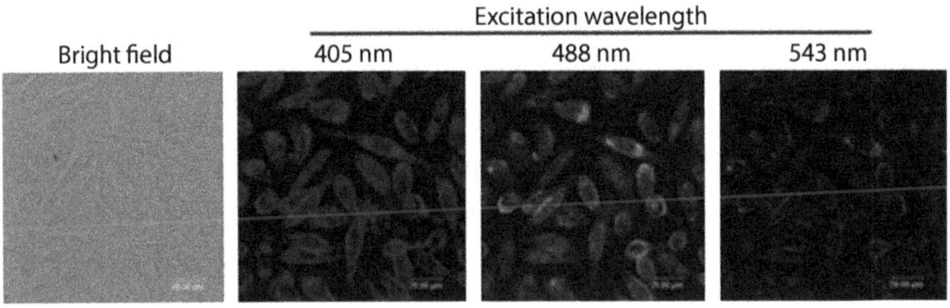

Figure 5.21 Laser scanning confocal microscopy images of ethylenediamine-carbon dots-labeled L929 cells. (Reproduced with permission from [347]; Copyright © 2012 Royal Society of Chemistry)

CD formation, thus confirming that an alkaline environment is the key factor and OH groups are essential for the formation of CDs by this electrochemical oxidation process.

According to work reported in 2012 by different authors, CDs have shown practical applicability for measuring *in-vivo* and *in-vitro* pH value. The CDs were synthesized from ascorbic acid (by direct heating), which exhibited a linear dependence on the pH of the solution in the range of 4.0 to 8.0 and had potential application in colorimetric and fluorescent probes for pH measurement [107]. Whereas, when CD was synthesized from L-asorbic acid as the carbon precursor, glycol and the triple distilled water as the solvent, this solvothermal approach resulted in green fluorescent CDs; this also showed a good linear relationship with the solution pH value in the range of 2.55–5.19 [336]. Another precursor has yielded pH-sensitive amino-coated CD by heating citric acid in glycerol at 220 °C for 3 h in the presence of 4,7,10-trioxa-1,13-tridecanediamine, then pH-sensitive FITC to pH-insensitive rhodamine B isothiocyanate (RBITC). The pH-dependent signal of dual-labeled CDs generated a good linear calibration curve in the pH range from 6.0 to 8.0 and the pH probe displayed an excellent reversibility between pH 5 and pH 9. Shi *et al.* [337] developed a tunable ratiometric pH sensor for intracellular pH based on the coating of CDs with amino groups followed by nanomaterial functionalization with fluorescein isothiocyanate and pH-insensitive rhodamine B isothiocyanate. The authors obtained real-time CD biosensors for intracellular pH in the range between pH 5 and 9, as demonstrated in the case of HeLa cells, where pH fluctuations associated with oxidative stress were successfully monitored.

Kong *et al.* [178] reported the use of CD-based inorganic-organic fluorescent probe for two-photon bioprobe for detecting physiological pH in tissues and in live cells. The CD was synthesized aminomethylphenylterpyridine (AE-TPY) CDs. The authors found that under one-photon and two-photo conditions, the fluorescent emission intensity showed good linearity with pH variation in the range of 5.4–8.6 and 6.0–8.5, respectively. Moreover, images of real-time changes in cytosolic pH were obtained in tumor cells upon exchanges of Na^+ and K^+. Interestingly, the authors demonstrated that CDs were able to monitor pH gradients up to 185 μm depth in living tissues, as revealed by two-photon microscopy.

Another precursor tried for synthesizing CD was threonine, using one-pot hydrothermal method Jin *et al.* [338]. The fluorescence of CDs was sensitive to pH in a broad range of 2.13–9.34, which was characterized by a decrease of the fluorescence intensity when the pH value increased. Under around physiological conditions (pH 6.03–8.91), the fluorescent emission intensity showed excellent linearity with pH variation. Moreover, the CDs were readily internalized into plant pathogenic fungal cells for imaging with dose dependence, and intracellular pH regulation was visually expressed as the variation of fluorescence intensity of the internalized CDs.

The use of CDs as biosensors of biological pH can also be achieved by treating the nanomaterial with acid reflux, leading to the formation of carboxyl and hydroxyl groups on the surface of CDs [339]. CDs were employed inside living pathogenic fungal cells to detect the intracellular pH value [267].

5.3.4 Detection of Nucleic Acid

Use of CDs in detection of DNA was first achieved in 2011 by Bai *et al.* [340]. They showed that methylene blue can effectively quench the fluorescence of CDs; hence, they joined ct-DNA in solution; the fluorescence of the solution was found to be restored. Based on this phenomenon, a new fluorescent sensor for DNA detection was designed; the detection limit was 1.0 × 10–6 mol/L with a linear range of 3.0×10^{-6} mol/L to 8.0×10^{-5} mol/L.

To detect nucleic acid, Li *et al.* [93] designed a CDs-based strategy for DNA sensing. The CD was obtained from candle soot and was labeled with single-stranded DNA (ssDNA), which quenched the PL of CD probe. When the combination of the target DNA and the labeled ssDNA formed double-stranded DNA (dsDNA), the PL was recovered when the labeled ssDNA gets disconnected from CDs. By this method the fluorescence signal could be detected and quantified upon hybridization of CDs with the target DNA, leading to the formation of double-stranded DNA (dsDNA).

Another method of DNA detection was developed by Wang *et al.* [214] using potassium permanganate to trigger chemiluminescence of CDs for DNA detection. The DNA detection was performed in a linear range from 10–18 to 10–14 mol/L with the low limit of detection being 8.56 × 10–19 mol/L. Owing to the sensitivity and rapid response of the CD-based sensor, this biosensor might be used to detect trace amounts of DNA in biological samples.

5.3.5 Detection of Vitamins

CDs-based sensors have received extensive interest for the detection of biomolecules like protein, amino acids and vitamins. Riboflavin and vitamin B12 have been detected by surface-functionalized CDs to target molecules [283, 169]. The ratiometric sensing protocol is widely accepted due to the fact that it eliminates the influence of temperature variation and local probe concentration. According to Yu *et al.* [330], this is mainly because the ratio of two interconnected PL signals is considered a detection index. The riboflavin is effectively detected by CDs-based ratiometric sensor with high sensitivity of 1.9 nM.

5.3.6 Detection of Protein and Enzymes

An aptamer-functionalized CD as a sensor for protein detection was developed by Xu *et al.* [155]. Since the thrombin has two different combining sites, the presence of thrombin could induce the CDs and aptamer to form a sandwich structure (Figure 5.10). The sensor shows high specificity and sensitivity towards thrombin. The detection limit of thrombin by this sensor reaches as low as 1 nmol.

DNA-labeled CDs have also been developed to detect protein and enzymes [341]. For instance, a novel CDs-dsDNA sensor was synthesized to detect histone [342]. In this sensor, the PL was quenched in the presence of dsDNA, while with the addition of histone, the luminescence was turned on through unwinding dsDNA from CDs due to the strong affinity between histone and dsDNA. With the binding affinity between DNA and proteins, the quantitative detection of protein can be calculated from fluorescence restoration.

A novel turn-on CDs-based sensor was fabricated to detect cysteine with excellent selectivity and sensitivity by Jana *et al.* [343] and a nanosensor composed of CDs and gold nanoparticles to detect cysteine with multiple signals was developed by Deng *et al.* [344].

5.3.7 Detection of Glucose

By now it is obvious that several molecules that are part of biological system can be successfully detected by CD-based sensors. Glucose is one of them. Wang *et al.* [282] developed CDs to detect glucose through peroxidase-like catalytic activity of CDs, which is achieved by assembling porphyrins on the surface of CDs by π-π stacking and electrostatic interactions. These CDs catalyze the appropriated peroxidase substrate, producing a blue color in the presence of H_2O_2. Moreover, the authors reported that the catalytic activity of modified CDs depended on the temperature and pH (optimal pH was found to be 4.0), and no inhibition of catalytic activity was reported even at high H_2O_2 concentrations.

Qin *et al.* [271] synthesized CDs by hydrothermal carbonization of waste of willow bark which showed 1–4 nm diameter and water solubility; photoluminescent CDs with quantum yield of approximately 6.0% was used as an effective photocatalyst. Then AuNPs/rGO nanocomposite was prepared by UV irradiation of the mixture of GO and $HAuCl_4$ aqueous solution in the presence of CDs produced from willow bark. It simultaneously reduced Au(III) complex and graphene oxide to form Au nanoparticles-decorated reduced graphene oxide (AuNPs/rGO) nanocomposite. This AuNPs/rGO nanocomposite can simultaneously reduce and oxidize H_2O_2 by catalytic action. Using the same nanocomposite, a glucose biosensor was fabricated by immobilizing glucose oxidase on the AuNPs/rGO-modified glassy carbon electrode for glucose detection. The linear response range and detection limit are estimated to be from 2 mM to 18 mM (*r*: 0.995) and 45 μM, respectively, when used as glucose sensor for human blood serum.

For immobilization of glucose oxidase as glucose biosensor, ionic liquid-functionalized CDs (IL-CDs), synthesized by electrochemical exfoliation of graphite rods in the presence of amino-terminated ionic liquid, have also been used as a matrix. The glucose oxidase immobilized CDs can detect glucose in the linear range from 20 μmol/L to 1 mmol/L with a detection limit of 7 μmol/L [345]. Therefore, CDs can be used for enzyme immobilization for several biomedical and technological applications.

5.3.8 Detection of Cancerous Cells

To better understand and detect the migration and proliferation mechanisms of cells, a means to track cells in living subjects is essential, and to achieve that, a number of cell labeling techniques and detecting probes have been developed. For sensing cancer cells [346] CDs were coated with three different polymers, $PEG_{1500}N$, triblock polymer PEIPEGPEI and 4-armed amine-terminated PEG (4-arm PEG); followed by the functionalization of CD with the protein transferrin through carbodiimide chemistry to target cancer cells. The authors reported an efficient internalization of PEG1500N and 4-arm PEG functionalized CDs in HeLa cells, indicating that the design of appropriated functional groups on the surface of CDs can be used to tailor the nanomaterial with cancer cells.

In another publication, mitochondrial H_2O_2 was successfully detected by a multifunctional CD-based fluorescent nanosensor by conjugating H_2O_2 recognition element and mitochondria-targeting ligand onto CDs. CD-based sensor was able to detect mitochondrial H_2O_2 in RAW 264.7 cells (RAW 264.7 cells are a macrophage-like, Abelson leukemia virus transformed cell line derived from BALB/c mice.) and in L929 cells with a limit of detection of 0.75 μmol/L [348].

5.3.9 Detection of Dopamine Neurotransmitter

In the brain, dopamine acts as a neurotransmitter (a chemical released by neurons) and plays a key role in the central nervous system; and outside the central nervous system (CNS) dopamine acts as a local chemical messenger. Neurodegenerative diseases like schizophrenia, Alzheimer's disease, and Parkinson's disease are related to abnormal levels of dopamine [349–351]. The dopamine precursor levodopa (*L*-3,4-dihydroxyphenylalanine, *L*-DOPA) is prescribed for the treatment of Parkinson's disease but has strong side effects (such as insomnia, loss of appetite, vomiting, hallucinations, etc.). Therefore, it is imperative to have a sensitive and efficient quantification probe for detecting even a trace level of *L*-DOPA in biological systems.

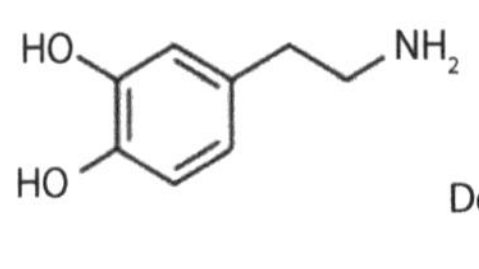

Dopamine (2-(3,4-Dihydroxyphenyl)ethylamine)

Li *et al.* [352] used a fluorescent CD coupled with tyrosine (TRY) for detecting *L*-DOPA, that showed a detection limit as low as 90 nmol/L and a linear range from 0.317 mmol/L to 31.1 μmol/L. When TRY was added to *L*-DOPA solution, the final mixture turned orange due to the formation of levodopa-quinone, which undergoes auto-oxidation to levodopachrome. Figure 5.22a shows the UV-Vis absorption spectrum of levodopachrome. Figure 5.22b presents the superposed graphs of the UV spectrum of levodopachrome (green line) and the photoluminescence spectrum of CDs (red line). The mechanism of *L*-DOPA detection by CDs proposed by them is presented in Figure 5.22c. Here, *L*-DOPA is converted to levodopachrome

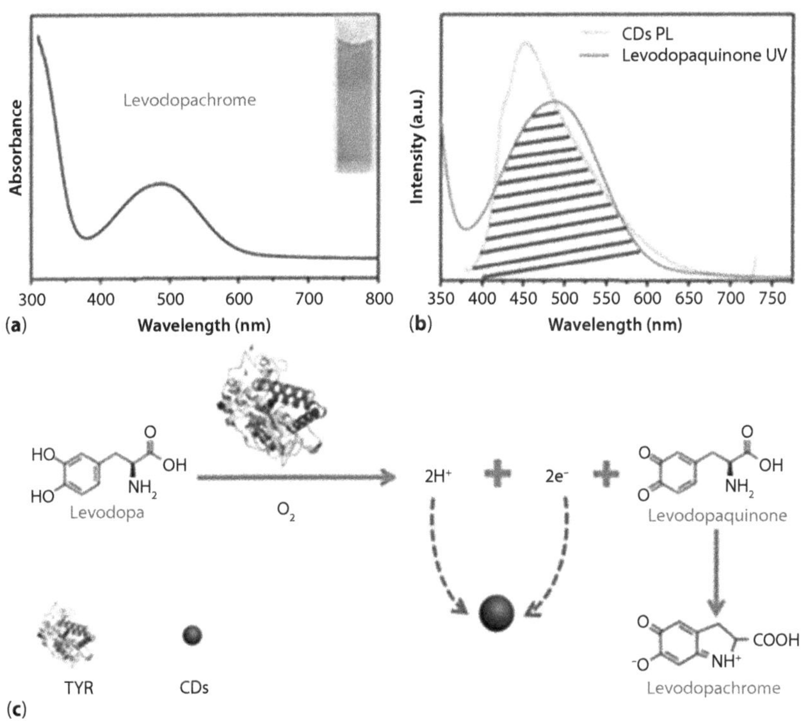

Figure 5.22 (a) UV-Vis absorption spectrum of levodopachrome; the inset is the corresponding digital photograph under visible light. (b) Superposed graphs of the UV spectrum of levodopachrome and photoluminescence spectrum of CDs. (c) Schematic illustration of the mechanism of detection of L-DOPA by CDs. (Reproduced with permission from [352]; Copyright © 2015 Royal Society of Chemistry.

by TRY, which has strong fluorescence absorption at 490 nm and is able to quench the fluorescence of CDs. The catalytic oxidation of *L*-DOPA produces levodopachrome, H^+, and e^-, shortening the reaction time.

Similarly, CDs obtained by green synthesis were assembled on a glassy carbon electrode for dopamine sensing by Li *et al.* [353], which had an even lower detection limit of 26 nmol/L. The sensor was employed for the determination of dopamine in human plasma, without ascorbic acid and uric acid interferences.

Simultaneously, Jiang *et al.* [354] prepared N-doped CDs (NCDs) by microwave-assisted technique. The obtained NCDs demonstrated selectivity and sensitivity to detect dopamine. A detection limit of much less than above two 1.2 nmol/L was obtained with this CD; with a linear range from 50 nmol/L to 8.0 μmol/L. These new NCDs have potential for directly and rapidly detecting the trace levels of dopamine in biological samples such as urine and human serum.

5.4 Concluding Remarks

As an emerging luminescent material, tremendous efforts have been devoted to employ CDs as a nontoxic alternative to replace traditional heavy metal-based QDs. In this chapter an overview of the bioapplication of CDs was presented. The importance of the physical and chemical properties attributed to bioapplications (high photostability and chemical stability, low cytotoxicity, and high quantum yield) were also touched upon. Then, bioapplications, such as biosensors and bioimaging, were systematically introduced. However, CDs prepared from diverse methods exhibit large size distribution and photoluminescence non-uniformity, and the complex and time-consuming separation and purification severely limit their further applications, impeding further bioapplications of CDs. Moreover, the mechanism of photoluminescence is still unclear. This demands that an effective synthesis method with a high yield should be developed to fabricate CDs with a high quantum yield in a small size distribution; and also an understanding of CDs' photoluminescence phenomenon, especially their bright multiphoton emission, should be explored to facilitate their *in-vivo* applications.

6

Suitability of Carbon Dots as Payload for Plants

Learn from yesterday, live for today, hope for tomorrow.
The important thing is not to stop questioning.
Albert Einstein

6.1 Introduction

Inputs of nanomaterials and QD in agriculture is still in its infancy but growing. As compared to scientists working in the field of various applications of nanotechnology in animals or human beings, plant- and agro-scientists have entered into the nano arena rather late. But that does not mean that there is any less scope or possibility of using nano or quantum dot material for the betterment of plants and agricultural crops. Already there are several reports based on the literature survey and patents that are looking into the application of nano-size material in the areas such as i) precision farming, ii) nanosensors, iii) food technology, and iv) nanofertilizers. So far as nanofertilizer research is concerned there have been efforts like incorporation of NPK fertilizer into CS-PMAA nanoparticles.

The question is why should we look at nanotechnology or carbon dots (CD) in agricultural or plant inputs, especially when it is hard to penetrate plant cell walls, the bark of trees, and the cuticular layer of epidermis of the leaves, stems and roots of crops, which are often covered with a waxy layer? It is because scientists are relying on the roots and leaves having nano- and micro-size pores through which nanomaterials can not only enter, but their large surface area to volume ratio can increase the interaction with the plant surface and improve the uptake [355]. Whenever the use of any material for the betterment of plants is considered, the agrosystem remains the main concern. The areas where CD can be integrated into plant/agro system are indicated in Figure 6.1. Use of nanotechnology for plants or in agrotechnology involves precision farming, nanosensors, food packaging and a less explored area, nanofertilizer. Because of limited arable lands and water resources, the development of the agriculture sector is only possible by increasing resource use efficiency with minimum damage to production

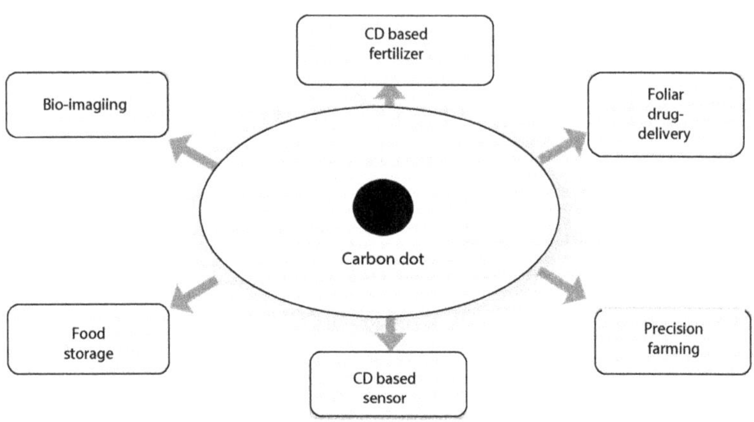

Figure 6.1 Possible applications and uses of carbon dots in agriculture.

beds through effective use of modern technologies. Among them, nanotechnology can revolutionize the agricultural systems. Nanotechnology has a high potential for achieving sustainable agriculture, especially in developing countries. Use of CD for plant system, especially as a nanofertilizer, is a relatively new and largely unexplored area of research that requires making a smart fertilizer, something able to deliver nutrients on demand to a crop. It sounds like drug delivery using nanoparticles for human therapeutics to a diseased tissue and not a healthy tissue; a very extensively researched field. In the same way, nutrients can be delivered to a crop when it needs it and not when it does not need it.

6.2 Suitability of Carbon Dots as a Payload for Plant

Before entering into the use of this nano-size carbon in plant system and the suitability of CDs as payload for fertilizer or nutrients, etc., it is important that we realize the importance of carbon in the soil, which is present in both inorganic as well as organic form. Usually the stable carbon that is present in the soil is the pyrolyzed biomass, or black carbon or biochar. The presence of these forms of soil offers improved soil productivity and biomass production. The large amount of carbon-based material in the "Amazonian dark earth" is responsible for their unusually high fertility over prolonged usage. The scanning and transmission electron microscopy, energy dispersive X-ray, electron energy-loss spectroscopy and Raman spectroscopy of such soil has shown that these carbon materials exhibit a complex morphology, with particles ranging in size from micrometers to nanometers [356]. These observations are indicative of the fact that carbon is a required material for soil and if they are applied as payload for fertilizer, chances are they will be useful.

The nano-size carbon material includes fullerenes, nano-onions, nano-cones, nano-horns, carbon dots (CD), carbon nanotubes (CNTs), nano-beads, nano-fibers, nano-diamonds, and graphene [357]. Nanoforms of carbon at the nanoscale < 10 nm combine the quantum effects, the stability of resonance structures and the tunable physicochemical properties due to the sp^3 character of C-C bonds [358].

Use of CD for developing nanofertilizer is among the emerging tools for sustainable agriculture. Studies have shown that the use of nanofertilizer causes an increase in nutrients use efficiency, reduces soil toxicity, minimizes the potential negative effects associated with overdosage and reduces the frequency of the application. In addition to this biogenically synthesized CD-based nanofertilizer being a complete biosource, it is eco-friendly and poses no health hazard. It is expected to be suitable for all crop varieties for improving soil aggregation, moisture retention and carbon buildup.

It has been recently suggested by Liu *et al.* [359] of Guangzhou University in China that carbon dots make good fertilizers. They describe a method to make a low-cost, biocompatible fertilizer with carbon QD derived from rapeseed pollen that increases the yield of hydroponically cultivated Rome lettuce by almost 50 percent.

In this chapter, our concern is CDs. One of the advantages of CDs is that over the last decade the cost-effective chemical and biogenic methods of CD synthesis has significantly evolved.

6.2.1 Cytotoxicity of Carbon Dots

The first necessary step in the use of any foreign material in any living system, be it plant or animal, is to make sure that the material is not toxic to the living system, unless it is used for killing them. At present, the growing interest in the use of CD is mostly confined to research level. However, there are concerns by the regulatory agencies and the researchers about their potential negative impact on the environment. Efforts are underway to study the ecotoxicity as well as cytotoxicity of nano forms of carbon materials including CDs, because as suggested by Jackson *et al.* [360], their unique and reactive properties may not only impact individual species but also disrupt ecological dynamics, when and if produced in very large quantities.

Fluorescent CDs are now being widely explored for their use in bioscience as payload for delivering nutrients, drugs and bioimaging. Hence, it is imperative that we know the effect of CDs on plants. Adopting mung bean as a model plant, which was incubated during both germination and growth periods, the phytotoxicity, uptake, and translocation of red emissive CDs in plants have shown that CDs at a concentration range from 0.1 to 1.0 mg/mL did induce some physiological response but did not impose any phytotoxicity or decreasing effect on the growth of the mung bean. The growth response was noted in terms of root and stem lengths. Treatment up to 0.4 mg/L exhibited an increasing trend. Concerning the uptake path of CDs, studies using confocal imaging showed that CDs were taken up by roots and transferred from the roots to the stems and finally to leaves by the vascular system through the apoplastic pathway [361].

Phytotoxicity of carbon nanomaterials are largely dependent on their concentrations, growth/exposure conditions, and plant species [362]. Moreover, functionalized MWCNTs (0–5000 mg/kg), having acidic nature, also affect the soil bacteria by decreasing the soil pH at higher exposure concentrations and subsequently temporarily affecting the soil bacterial communities. But after some time, this inhibitory effect does not persist. Therefore, while using CD very many parameters, like acidic nature, dose concentration and duration, etc., should be considered.

6.2.2 Carbon Dots and Plant Growth

It has been already shown that a sister nanoparticle of CD, carbon nanotubes (CNT), i) enhances the *in-vitro* growth of seedlings [363] by penetrating the seed coat (Figure 6.2); ii) acts as growth regulator in tomato [364]; iii) induces enhancement

Figure 6.2 Tomato seedlings germinated *in vitro* with (in black culture media) and without CNT in the culture media. (Reproduced from [371])

of cell growth of tobacco [365]; iv) enhances germination and growth in *Glycine max* [366]; v) enhances root elongation in crops [367]; vi) enhances growth of gram [368]; and vi) enhances growth of wheat [369, 370]).

Though there are reports of nanoforms of carbonaceous material enhancing the growth, there has not been any clear indication as to how or what is the *modus operandi* of growth enhancement. However, Khodakovskaya *et al.* [363] have suggested that CNTs penetrate thick seed coat of tomato and support water uptake inside seeds, thus affecting seed germination and growth of tomato seedlings. But at this stage it is hard to say whether it is good news or bad to use CNT; because using CNT as fertilizer could have unintended consequences and effects on the environment, human beings or other organisms through the food chain, which aren't very well understood or studied.

This growth promoting effect of CNTs has given a directive that nano-sized CDs, which are much smaller than CNT, may be an even better growth enhancer than CNT. Hence, the influence of water-soluble CD on the growth of various plants has been recorded. Moreover, CDs can be more easily metabolized or removed from the system than CNTs. Some crops that have shown enhanced growth due to application of CDs are mentioned below.

6.2.2.1 Wheat

Wheat plants exposed to 150 mg/L (200 μl) water-soluble CD for 10 days exhibited ten times the enhanced root and shoot growth [372] as compared to control (Figure 6.3).

6.2.2.2 Green Beans

Green beans were germinated in a petri dish in aqueous solution of water-soluble N-doped CD of 0.1 to 10.0 mg/L concentration (Figure 6.4). Control green beans were germinated in pure water. To prepare N-doped CD a microwave-assisted method was used through heterogeneous reaction between calcium citrate (insoluble solid as source of carbon) and urea

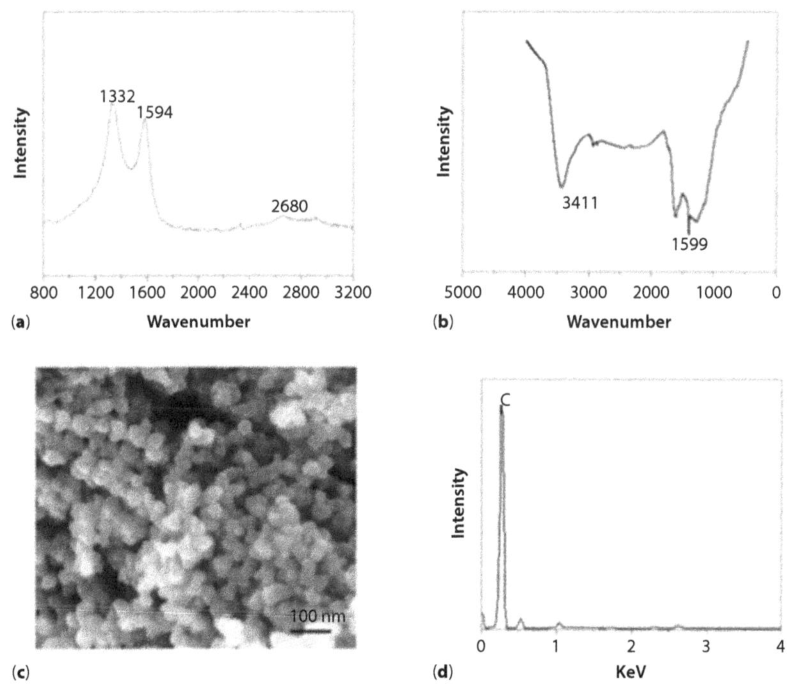

Figure 6.3 (a) Raman spectra of carbon nanodots (CND) showing the characteristic D and G band; (b) FTIR of the water-soluble CND; (c) SEM images of water-soluble CND; (d) EDX of CND showing the major presence of carbon and small amount of oxygen. (Reproduced with permission from [372]; Copyright © 2014 Springer)

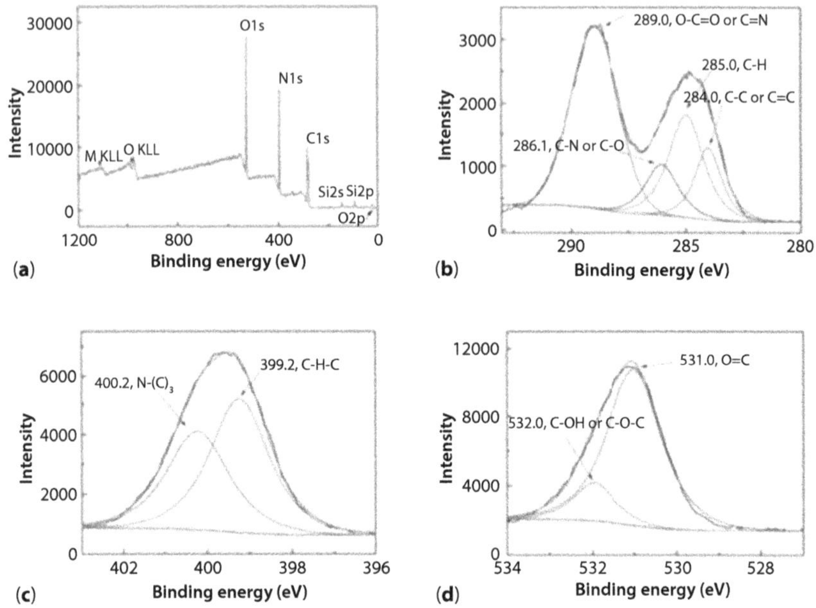

Figure 6.4 XPS survey scan of N-doped CDs. (a) XPS scanning spectra show three major peaks of carbon, nitrogen and oxygen. XPS high resolution survey scan of (b) C1s, (c) N1s and (d) O1s region of N-doped CDs. (Reproduced from [70])

(as nitrogen source for doing CD) at 800 W for 5 min, which yielded yellowish-green fluorescence. Urea may be ascribed to a surface state in which urea groups form intramolecular H-bonds [373].

As shown in Figure 6.5, the fluorescence of bean sprouts indicated permeation and distribution of N-doped CDs throughout the plant cells. Moreover, they were biocompatible, nontoxic and do not hinder growth of green bean seedling. However, the growth was observed to be concentration dependent.

6.2.3 Is Cell Wall a Barrier for Carbon Dot Internalization?

To enter a cell or act as a payload to deliver nutrients or drugs to the plants CD has to face two barriers: cell wall and cell membrane. Membrane transport mechanism is a very well understood phenomenon, as it has allowed various nanoparticles and

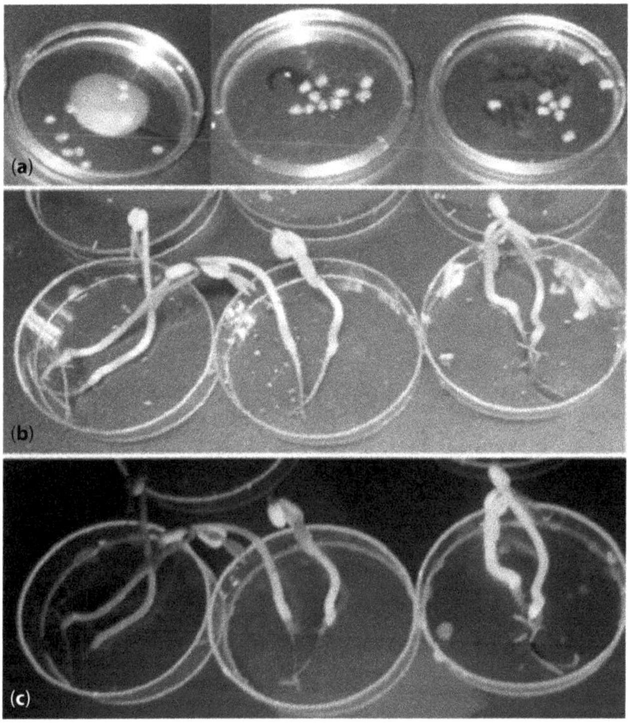

Figure 6.5 (a) Green beans were added into solution; bean sprouts photo under (b) daylight and (c) 365 nm UV beam after a few days growth. From left to right: pure water, N-doped CDs solution (our method), N-doped CDs solution (citric acid and urea). (Reproduced from [373])

QD to get internalized by endocytic vesicles and the lumen of the organelles supports secretory pathways. Uptake through cell membrane is mostly active.

To understand the uptake of nutrients through the plant cell, let us first look at the basic structure of a plant cell wall (Figure 6.6). The cell wall is a rigid, semipermeable protective layer composed mainly of cellulose layer positioned next to the cell membrane in plant algae, fungi, some microbes and archaea. However, animals do not have a cell wall. The plant cell wall consists of three layers: i) *Middle lamella* **is** the outermost layer of the cell wall **composed of** pectin, which is a polysaccharide that aids in cell adhesion by binding with the adjacent

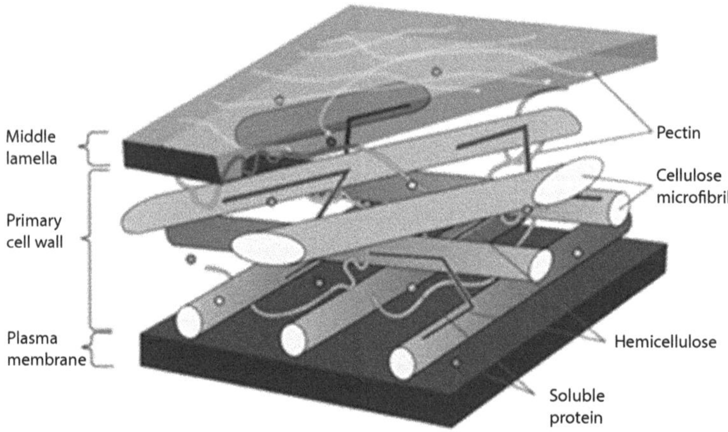

Figure 6.6 Schematic of cell wall structure.

cell wall; ii) the next layer is called the *Primary cell wall*, which is between the middle lamella and cell membrane, and is composed of cellulose microfibrils contained within a gel-like matrix of hemicellulose fibers and pectin polysaccharides. It provides the strength and flexibility needed for cell growth; and finally iii) *Secondary cell wall* (all plants do not have a secondary cell wall) is a layer formed between the primary cell wall and plasma membrane in some plant cells. When the primary cell wall stops dividing and growing, it starts thickening and is composed of cellulose and hemicelluloses and lignin. This layer gives strength to the plants and aids in water conductivity through the vascular system.

In spite of having three layers, a plant cell wall is penetrable because it has holes in the wall through which a cytoplasmic fibril called *plasmodesma* passes, which connects neighboring cells and allow the transfer of nutrients, waste, and ions (symplastic pathways). Molecules can also pass through the spaces within the cell walls, avoiding the cells completely (apoplastic pathways) (Figure 6.7).

Berestovsky [374] has suggested that plant cell walls have a system of pores that are <10 nm in diameter. The plant cell wall, in addition to its mechanical function, also functions as a

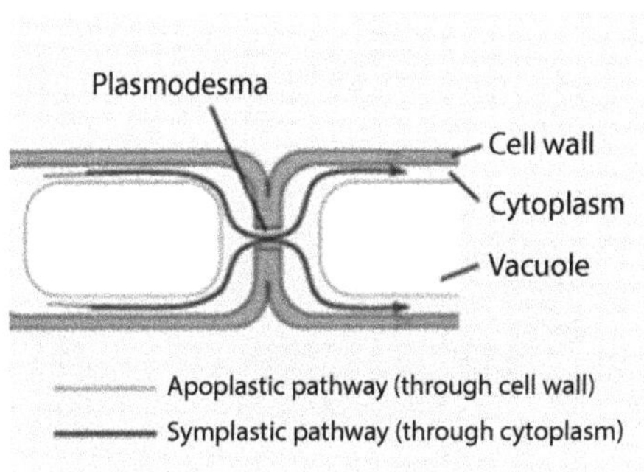

Figure 6.7 Schematic representation of plasmodesma.

porous network, determining the upper size limit for the molecules which the cell may exchange with its external medium. The limit is determined by the diameter of through pores in the cell wall. Despite the importance of the size of those pores, very little information is available concerning the pore diameters in the cell walls of plants.

These structures of cell wall and their functions suggests that nutrients move from cell to cell and so cell wall will not be a barrier for transport of nutrients supplied using CD as payload. However, the problem of losing water through the holes in the middle of the day or on very hot days is a consideration to be noted. During daytime plants lose water through the stomatal pores by a process of transpiration. Therefore, nutrient uptake should take place during early morning or after sunset.

6.2.4 Other Possible Routes for Entry of Carbon Dots

Roots and leaves have nano- and micropores; hence, the small size and large surface area of CDs can increase the interaction with plant surfaces, perhaps improving uptake (Figure 6.8).

Let's look at how plants obtain their required nutrients from the fertilizer added to the soil and the role of carbonaceous material. Humus is known to be found in soil, especially black soil; it is a colloidal carbonaceous residue that serves as a nutrient reservoir. Plants take up the nutrients from the soil by exchanging cations, where root hairs pump out hydrogen ions (H^+)

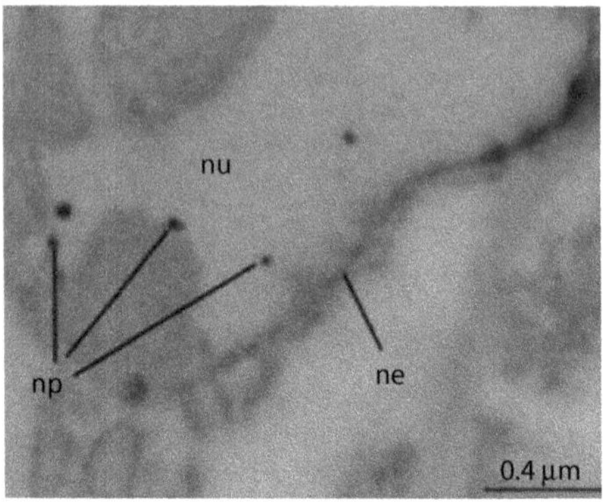

Figure 6.8 TEM image of nanoparticles taken up into ryegrass root (Reproduced from [375])

into the soil through proton pumps in which hydrogen ions displace cations attached to negatively charged soil particles so that the cations are available for uptake by the root.

Root hairs are involved in the uptake of nutrients; after nutrient uptake root hairs transport them to the center of the root so that the nutrient reaches the conducting tissue *xylem* and *phloem*. These conducting tissues or the vascular bundles are surrounded by a cell wall known as the *Casparian strip*, which prevents passive flow of water and nutrients so as to regulate the nutrient uptake. Xylem is involved in the transport of water and inorganic molecules in plants and phloem translocates organic molecules. Water potential is very important here. If the water potential is more negative within the plant than the surrounding soils, the nutrients will move from the higher solute (soil) concentration to lower solute concentration (plant). Nutrients are taken up through the roots by i) Simple diffusion, which occurs when nonpolar molecules (O_2, CO_2, and NH_3, etc.) that follow a concentration gradient, can passively move through the lipid bilayer membrane without the use of transport proteins; ii) Facilitated diffusion, which involves rapid movement of solutes or ions following a concentration gradient, facilitated by transport proteins; iii) Active transport, which requires an energy source, usually ATP, to pump the ions or molecules through the membrane against a concentration gradient.

6.3 Carbon Dots and Plant Fertilizer

Application of nanotechnology in agriculture is in the nascent stage. But the recent trends have shown a change in approaches through nanotechnology. The rationale of using nanotechnology has directed the agro-scientists to consider a need to design nanoparticles having enhanced biocompatibility to develop small enough particles to penetrate through the surface pores of leaves and roots; unlike conventional modes of fertilizer delivery, it should offer controlled release of fertilizer to decrease the unwanted burden on the soil such as increased acidity due to surplus nitrogen. Moreover, nanoparticles loaded with more than one fertilizer, viz. NPK, to enhance their synergistic effect should also be pondered. One of the issues that is plaguing agriculture today is the lack of efficiency of fertilizers. Fifty to seventy per cent of fertilizer—nitrogen, for example—applied to crops is actually not taken up by the plants at all. It is lost to the air, to water and other processes. All this wasted nitrogen is not only a severe economic issue for farmers, as it is not going into their crop yields, but is also an environmental problem.

Though attention is being paid to nanoscale fertilizer and supplement particles, disadvantages like more loss to leaching and microbes may be a possibility and should not be overlooked. One more thing should be considered, and that is the fact that fertilizer is not nanostructured. Though there are some existing commercial nanofertilizer products, none of them uses CDs.

Successful use of CD prepared from the pollen of rapeseed has been achieved as carrier of nutrients for hydroponically cultivated *Lactuca sativa* (lettuce). This CD has good water solubility and biocompatibility. More than 50% enhanced growth of plants treated with nutrients loaded along with CD were recorded using growth indices like morphology, nutrition quality of the plant, gas exchange, and photosynthetic pigment content [376]. The assessment of nutrient quality showed that the composition of vitamin C, soluble sugars and proteins in the lettuce were not altered by the carbon dots, declaring it a safe method to increase crop yields. Moreover, the transport route of CD was also evaluated under UV light excitation. It was found that the carbon dots were mainly accumulated in the leaves. Since they added CD to Hoagland solution used for hydroponic cultivation, it is difficult to say if CDs are carrier of fertilizer. However, it cannot be denied that these pollen-derived CDs acted as a good fertilizer, as they did promote plant growth. The increase in 50% plant growth was obtained when 30 mg of CD/l nutrient solution was applied as compared with plants that did not receive the carbon dots.

6.3.1 Nitrogen

Nitrogen is one of the key nutrients for plants. It is required for the synthesis of biomass, food, wood, primary and secondary metabolites and fibers. Hence, nitrogen is the most required and important fertilizer. Conventional nitrogen fertilizer is applied in more than the required amount for plants. This is because more than 50% water-soluble nitrogen applied as nitrate is lost to the soil due to leaching; a good amount of loss is by emission as ammonia and nitrogen oxide and soil microorganism-mediated is incorporated into soil organic matter over time [377]. Usually urea is used as a rich source of nitrogen, as fertilizer. The problem is that it breaks down quickly in wet soil and forms ammonia, which is washed away, creating an environmental concern as it leads to eutrophication of waterways and finally enters the atmosphere as nitrogen dioxide, the main greenhouse gas associated with agriculture. Therefore, more fertilizer needs to be applied to boost production.

One of the concerns in applying nitrogen to plants is that the nitrogen utilization efficiency (NUE) by plants is low. This demands an effort to increase the NUE in conventional fertilizer formulations, but so far has not been very successful. Nanotechnologists feel that due to the high surface area to volume ratio, nanofertilizers may be more effective than polymer-coated conventional slow-release fertilizers [198, 378].

Nanoscale fertilizer of reduced size could make it easier to dissolve and more potent. Therefore, trials are being done by Kottegoda *et al.* [379] that use nano strategy for slow release fertilizer based on urea-modified hydroxyapatite (HA) nanoparticles encapsulated into the cavities present in soft wood of *Gliricidia sepium*. They studied the release behavior of nanofertilizer against common fertilizer at three different pHs, 4.2, 5.2 and 7. The nanofertilizer showed an initial burst and subsequent slow release even on day 60 compared to the commercial fertilizer, which had an early heavy release followed by a release of low and non-uniform quantities until around day 30. Nanoscale fertilizer of reduced size could make it easier to dissolve and more potent. However, more loss to microbes needs to be considered when nanofertilizers are used.

Major considerations in supplying nitrogen to plants using CD – Nitrogen applied to the soil is higher than required, 50–70% of fertilizer applied to crops is lost to air, water and other processes and it is an economic issue for both farmers and the environment.

Attention is being paid to nanoscale fertilizer and supplement particles because:

i. It reduces the size of existing chemicals to the nanoscale.
ii. Due to their high surface area relative to the amount of nanomaterials, fertilizers based on nanotechnology have the potential to surpass conventional fertilizers. Smaller size could mean more potency. Hence, because of the high surface area relative to the amount of nanomaterials, fertilizers based on nanotechnology have the potential to surpass conventional fertilizers.
iii. Roots and leaves have nano- and micropores. The small size and large surface area of nanomaterials can increase the interaction with the surfaces of roots and leaves having nano- or microsized

pores and may improve the uptake of desired molecules.

iv. Nanotechnology can impact the reduction of spray drift with aerodynamic nanoparticles for optimized deposition and controlled release for season-long treatment.

v. CDs are water soluble. However, disadvantages like more loss to leaching and microbes may be a possibility and should not be overlooked. One more thing should be considered, and that is that fertilizer is not nanostructured. Though there are some existing commercial nanofertilizer products, none of them uses CDs.

Another consideration is in which form nitrogen should be supplied as nanoparticulate system. This will depend on whether:

i. It is applied through foliar spray. Then the soluble forms of nitrogen that are transported in the plants as amines and amides (foliar application) may be successful.

ii. If these soluble nano-forms are applied through the soil then application of CD-conjugated nitrogen as nitrate with targeting molecules for higher uptake should be tried.

iii. Or both the above applications can be combined.

6.3.2 Phosphorus

Phosphorus (P) is an essential nutrient for all living organisms. Plants need phosphorus for normal growth and maturity because it plays an important role in photosynthesis, respiration, energy storage and transfer, cell division, cell enlargement and several other processes. In soil all phosphates are not available for the plants to absorb them because they are present in water-insoluble form. Phosphate-solubilizing bacteria present

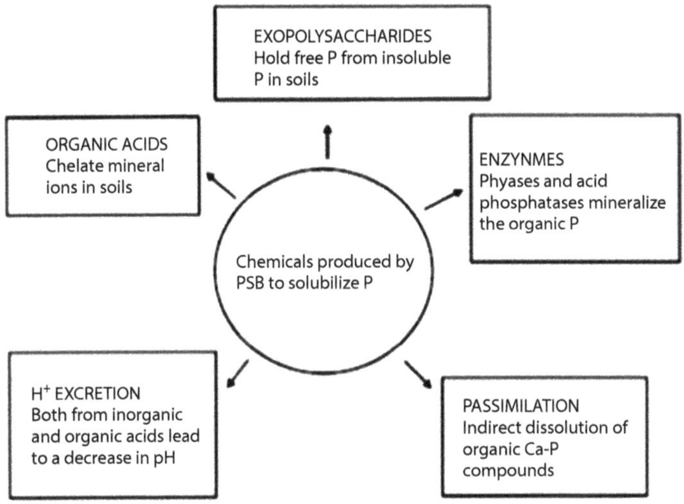

Figure 6.9 Mechanisms of phosphorus (P) solubilization by phosphate-solubilizing bacteria.

in the soil are needed for the uptake of P (Figure 6.9). Along with nitrogen, phosphorus, and potassium (NPK), other nutrients are a must for plant growth. But plants don't get the chance to use more than half of the nutrients in fertilizer because 80–90% of the phosphorus compounds escape via leaching. Therefore, smart fertilizers are needed that will release phosphorus only at the time and in the amount needed and preferably with capacity for slow release. P-enriched biochar with nanotechnology has been tried by Bin Gao's group [380]. Biochar is charcoal produced from plant organic matter and when it is added to the soil it removes carbon dioxide from the air and fertilizes the crop by increasing the pH of the soil, which leads to improved availability of phosphorus and potassium [381]. CD, a carbonaceous material, may give the same additional advantage as biochar.

6.3.3 Potassium

In general, it can be said that the uptake of a certain N form closely matches the growth-related demand of the plant, at least when potassium (K) transport to the root surface is not limiting. In addition, many plants accumulate large pools of N and K during vegetative growth which is remobilized during the reproductive stage. As a consequence, systems responsible for N and K transport need to be tightly regulated in their expression and activity upon sensing that their graphene quantum dot has already been synthesized [382]. It would be interesting to see whether this type of conjugate can be used for controlled delivery of K to the plants.

Like N and P, most parts of K fertilizers are also lost as runoff or volatilized, i.e., approximately 50–70% of potassium of the applied fertilizers is lost to the environment and can't be absorbed by applying to the plant. Though many natural and synthetic biodegradable polymers, such as chitosan nanoparticles (~78 nm), have been used for the sustained release of fertilizers, now it is felt that CD may also have potential application in the slow-release behavior of fertilizers that needs to be researched.

6.3.4 Micronutrients

Micronutrients like manganese, copper, boron, iron, molybdenum, zinc, etc., are essential for the growth and development of plants. They are often applied through the foliar application of micronutrients and can enhance uptake by the leaves [383]. CD-attached micronutrient formulations can be used to make micronutrients available to plants. They can be sprayed on plants or even supplied to soil for uptake by roots to enhance soil health and vigor. Different nanoparticles of micronutrients as such have been tested to provide an appropriate level of micronutrients to plants.

Carbon nanofibers, a member of carbon nanoforms, have been used as a micronutrient carrier in plants and could efficiently translocate and control release of the nanoparticles of copper (Cu) and zinc (Zn) [384, 385]. This gives an indication that CD may also be a good carrier of micronutrients.

Apart from fertilizers, herbicides, pesticides, etc., can also be attached to the CD for controlled and precise release. For successful use of CD as NPK fertilizer, selection/screening of NPK-rich compounds that can be used for attachment to CD and optimization protocol for designing nanovehicle, as well as entrapment chemistry of fertilizer, should be understood. Preliminary research in this direction needs understanding of the release kinetics of fertilizers under physiological and anatomical conditions of plants and the environment. A study of the impact of synthesized nanocomposites on model plants in terms of uptake, controlled release by plants and toxicity, if any, are also of utmost importance.

6.4 Need for Sensor to Detect

Sensors are part of precision farming that helps in maximizing the output from crops while minimizing the input of fertilizers, pesticides, herbicides, etc., through specific sensors to monitor variable parameters and accordingly applying the action. Nowadays, nanosensors are being developed, as it has shown higher specificity and effectivity. The use of sensor involves computers, global satellite positioning systems and remote sensing devices to measure crop growth and its physiological status so as to make better decisions regarding inputs to the crops. Nanosensors combine the knowledge of biology, chemistry and nanotechnology and are also referred to as nanobiosensors. Nanosensors have the same arrangement as ordinary sensors but are produced at the nanoscale. Hence, they are extremely small devices which detect and respond to physicochemical and biological signals. Nanosensors-supported fertigation and CD-based smart delivery systems are bound to help in the efficient use of natural resources like water, nutrients and agrochemicals by precision farming [386].

Due to more efficient luminescence, small emission spectra, excellent photostability and tenability according to the particle sizes and material composition, CDs can be far better and rapid than organic fluorescent dyes. By a single excitation light source, CDs can be excited to different colors due to their broad absorption spectra. Already there are nanosensors to detect contaminants, pests, nutrient content, and plant stress due to drought, temperature, or pressure. Use of CD for preparation of different kinds of biosensors would be useful in remote sensing devices required for precision farming, which are part of the boon to this modern nanotechnology.

Different methods can be used to prepare CDs that can emit different colored photoluminescence, e.g., deep UV, blue, green, yellow, etc. However, blue and green PL of CDs are the most common ones. The PL of CD depends on many factors such as size, shape, excitation wavelength, pH, concentration, surface oxidation degree, surface functionalization, N-doping and S-doping. Though the PL mechanism of CDs remains controversial, generally it is derived from intrinsic state emission and defect state emission. Due to the photoluminescence properties of CDs, they have been exploited to detect many analytes.

6.4.1 Plant Nutrient Status

Detection of plant nutrient status in plants, water and soil will help in deciding the dose of fertilizer as well as time of application of fertilizer. Due to their fluorescence, CDs have excellent optical properties, chemical stability and good water solubility and have been used for chemical sensing, i.e., metal ion detection, anion detection, small organic molecules, and biomolecules detection. Like other semiconductor QD, by interacting with the analyte, CDs can change the efficiency of recombination between the surfaces of the electron-hole pairs which occurred in the fluorescence enhancement and quenching treatment to achieve quantitative or qualitative analysis of the measured object. Scientists have developed a variety of methods to detect plant micronutrients like Cu^{2+} [116, 387], Fe^{3+} [48, 321], etc.

Various CD-based sensors have been developed for human system and water analysis such as sensor for detection of variation in pH [388], chromium [Cr(VI)] in water [389], Cu^{2+} [318], mercury(II) ions [292]. Apart from metal ions, CD-based probes have also been developed for determination of organic molecules such as nucleic acid [390], Vitamin B [391], Glutathione [392], etc. A similar system can be modified and applied for the plant system. Bandi *et al.* [393] have reported CD synthesized from onion peel waste that has 28% quantum yield and exhibited high aqueous dispersibility, excitation-dependent fluorescence emission and excellent stability to various effects like pH, high ionic strength and continuous irradiation. Since the presence of Fe^{3+} ions could result in a strong fluorescence quenching, these CDs were used as a fluorescent probe for the selective and sensitive detection of Fe^{3+} ions and a good linear correlation (R2 ¼ 0.996) in the range of 0–20 mM with a detection limit of 0.31 mM was obtained. This sensor could also be used for plant, soil and water system. N-doped GQD fluorescent probe has also been prepared to detect label-free Fe(III) in aqueous medium [394].

Zong *et al.* [319] have shown that their specially fabricated CDs can bond with Cu^{2+} specificity and quench the fluorescence of CDs. Moreover, they could retrieve the bonded Cu^{2+} by adding L-cysteine in the solution of CDs-Cu^{2+}; the Cu^{2+} gets released from the surface of the CDs, thereby recovering the fluorescence of CDs also. With these chemical sensors, which successfully detected two substances—Cu^{2+} and L-cysteine—the detection limit of Cu^{2+} was 2.3×10^{-8} mol/L and of L-cysteine was 3.4×10^{-10} mol/L.

6.4.2 Water

The importance of water for a living system, be it plants, animals, microbes or algae, cannot be emphasized enough. Plants need water at all the stages of growth right from germination to fruit ripening. Soil water is the main source of water for plants.

When soil dries out, plant growth slows down, photosynthetic activity is reduced and the plant starts to wilt.

MIT engineers, Koman *et al.* [395], have created nanosensors that can be printed onto plant leaves and reveal when the plants need water by monitoring the stomatal function in terms of plant hydraulics, photosensitivity, and gas exchange (Figure 6.10). For this purpose a nanoparticle-based conducting ink that preserves stomatal function to print a highly stable, electrical conductometric sensor actuated by the stomata pore itself is repeatedly and reversibly utilized for over 1 week. As explained by them, "This stomatal electro-mechanical pore size sensor (SEMPSS) allows for real-time tracking of the latency of single stomatal opening and closing times *in planta*, which we show vary from 7.0 ± 0.5 to 25.0 ± 0.5 min for the former and from 53.0 ± 0.5 to 45.0 ± 0.5 min for the latter in *Spathiphyllum wallisii*. These values are shown to correlate with the soil water potential and the onset of the wilting response, in quantitative agreement with a dynamic mathematical model of stomatal function. A single stoma of *Spathiphyllum wallisii* is shown to distinguish between incident light intensities (up to 12 mW cm^{-2}) with temporal latency as slow as 7.0 ± 0.5 min. Over a seven day period, the latency in opening and closing times are

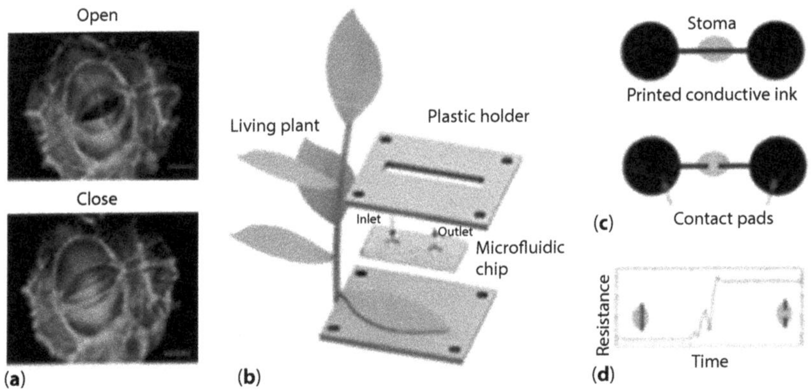

Figure 6.10 Graphical illustration of monitoring plant water status using a microfluidic-printed electro-mechanical sensor of stomata. (Reproduced from [395])

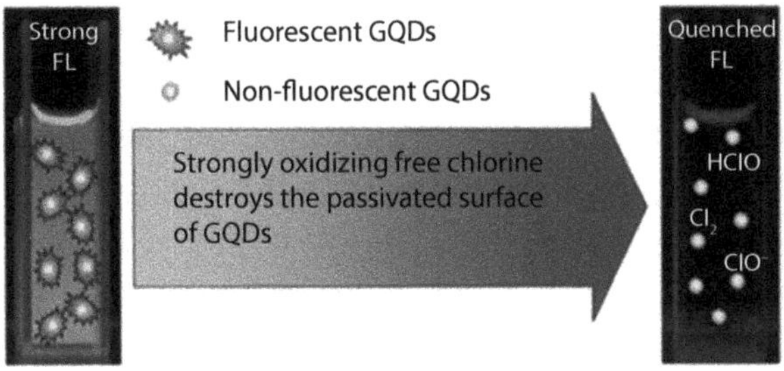

Figure 6.11 Graphical illustration of monitoring free chlorine in water. (Reproduced from [313])

stable throughout the plant diurnal cycle and increase gradually with the onset of drought."

A high amount of chlorine in irrigation or drinking water is not healthy for either. Graphene QD (GQDs) which possesses similar characteristics as CD, i.e., both are OD material with the quantum confinement and edge, have found application in sensor fabrication and address the advantages for developing smart sensors (Figure 6.11). Zong *et al.* [319] found that free chlorine destroys the passivated surface of GQD, resulting in fluorescence quenching signal. For fabricating this sensor, GQD is synthesized from citric acid by optimizing the parameters such as pH of solution, concentrations of GQD, and response time. The linear response range of free chlorine in water ($R^2 = 0.992$) was from 0.05 to 10 μM. This detection limit is much lower than the DPD (*N*-*N*-diethyl-*p*-phenylenediamine) colorimetric method. GQD has also been tried to detect CO_2 dissolved in water by Raeyani *et al.* [396].

6.4.3 Pathological Status of Plant

At present the detection technique for plant diseases takes days to find the plant disease; therefore, efforts are underway to find a faster, simpler portable detection system using nanotechnology, so that the combined efforts could lead to a better detection system. One of the problems with plants is that unlike humans they do not have an immune system and if any new plant pathogen is present in the area, plants are not able to protect themselves. Another problem is that if gene therapy is tried, then the nanoparticles that can penetrate plant cell walls simultaneously deliver a gene and a chemical that triggers its expression with controlled precision.

Apart from CDs other QD have been used to detect pathogens associated with different plant diseases. Nanotechnology-based plant viral disease detection kits are available and are being used in speedy and early detection of viral and other plant diseases. For example, QD-fluorescence resonance energy transfer-based sensors have been developed by Rad *et al.* [397] to detect witches' broom disease of lime caused by *Phytoplasma aurantifolia* and *Rhizomania* disease of sugar beet caused by beet necrotic yellow vein virus and the *Polymyxa betae* Keskin. The vector of beet necrotic yellow vein virus has been detected by QD-fluorescence resonance energy transfer-based sensor by Safarpour *et al.* [398]. QD are known to be toxic to the living system and, therefore, CD-based detection system has to be worked out.

6.4.4 Residual Pesticide, Herbicide or Insecticide in Plants

For residual chemical detection, QDs of cadmium telluride (CdTe) and H cadmium telluride (CdTe) have also been used. Zheng *et al.* [399] developed a highly sensitive organophosphorus pesticide (in vegetables and fruits) biosensors based on nanostructured films of acetylcholinesterase and CdTe QD. This biosensor is based on enzyme inhibition mechanism with detection limits of 1.05 × 10–11 M for paraoxon and 4.47×10^{-12} M for parathion, which is better than the conventional gas chromatographic methods or amperometric biosensors (0.5 nM). Thiol-stabilized luminescent cadmium telluride (CdTe) QD has also been used to detect smethyl parathion at picogram levels by Chouhan *et al.* [400]. Moreover, water-soluble CdTe QD and highly fluorescent silica molecularly imprinted nanospheres embedded cadmium telluride QD have been used as a biosensor for determination of deltamethrin in fruit and vegetable samples [401]. Here also the use of CD-based sensors remains a much needed unexplored area.

6.5 Carbon Dots and Foliar Application of Drugs on Plants

As mentioned in the previous chapter, a cell wall covering the innermost plasma membrane layer is a major consideration when thinking about uptake of any molecule by plants. Let us take a cursory glance at various layers of the plant cell wall starting with the outermost layer:

6.5.1 Cuticular Layer

The cuticle is an important layer as it protects plants by controlling water loss and protects from UV radiation. The thickness of this layer varies according to the plant from 1–10 μm thick and has water-filled pores. The cuticle covering the *stomata* (Figure 6.12) and *trichomes* forms a site for *foliar penetration.* The cuticular layer is composed of cutin, wax, pectin and cutan. The lipophilic cutin and wax forms the largest fraction of the cuticle. But there are polar domains also present for penetration of polar compounds. Cutin is a de-polymerizable polymer

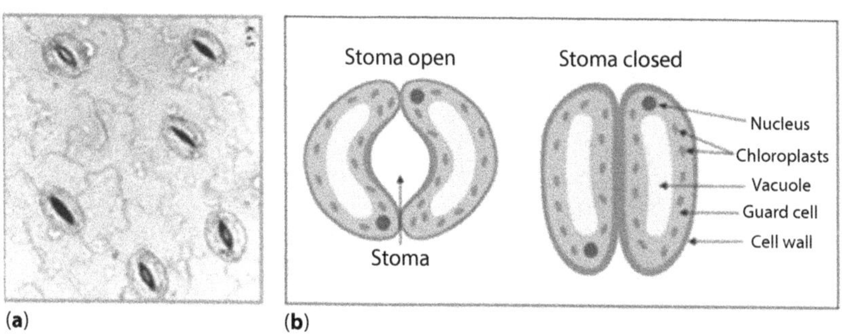

Figure 6.12 (a) Epidermal cuticular layer of a leaf showing open stomata and (b) schematic diagram showing details of the stomata.

that is deposited as 2-dimensional polymer membrane on the surface of above-ground plant parts. Some small polar but uncharged molecules (water), such as ionic compounds (Ca^{2+} salts and glyphosate as an organic charged molecule) also have cuticular permeability and they accumulate in the guard cells of the stomata and base of trichomes (Figure 6.13).

A wax layer is found on the cuticle surface as intracuticular wax. The wax layer contains soluble cuticular lipids having linear and long hydrocarbon chain aliphatic molecules with different functionalities (alkanes, alcohols, aldehydes, acids, etc.). They form solid, partially crystalline aggregates at room temperature.

Structural barriers to the penetration and uptake of antibiotics, nutrients, herbicides, pesticides or PGRs depend on both structural and physiological parameters. For example, trichomes reduce spray droplets from reaching the leaf surface, whereas volatile materials can easily penetrate through the stomata; but aqueous solutes do not penetrate easily unless the surface tension of the spray solution is decreased by using a surfactant. In polymeric cell wall materials, such as cellulose, lignin, etc., foliar penetration depends on parameters such as:

i. Molecular size of ions – A four-fold increase in molecular weight decreases the mobility through

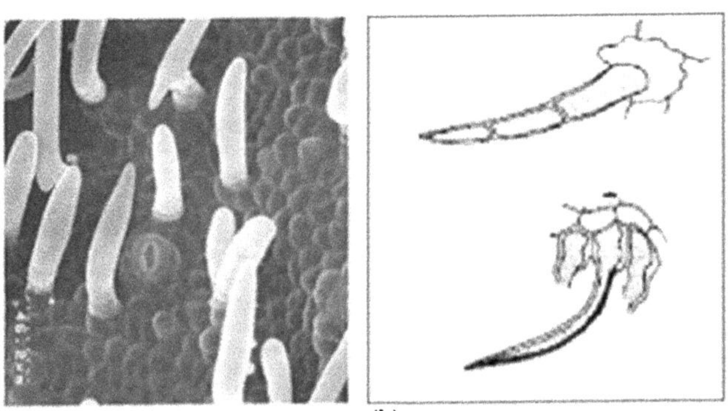

(a) (b)

Figure 6.13 (a) Epidermal layer of a leaf showing trichomes and (b) schematic diagram showing details of trichome.

the cuticle by a factor of as much as 1000. This confirms possible diffusion across the wax.

ii. Humidity – Cuticular transpiration is affected by humidity. Cuticular transpiration is correlated with ion permeability to lipophilic molecules. Water uses the same transport path across the cuticle. An increase in atmospheric humidity from 0 to 100% causes a decrease in transpiration by a factor of 2 to 3. This means that water molecule is absorbed by the cuticle.

iii. Plasticizer – Mobility can be increased by adding lipophilic accelerator compounds, because they dissolve wax in cutin and act as plasticizer by decreasing the barrier property.

A study of these parameters shows that ions penetrate cuticles via water-filled pores.

6.5.2 Cell Wall

Porous cell wall composed of cellulose is the next barrier after the cuticular layer. Both polar as well as nonpolar molecules can move through it, but compounds with ~700 mol wt do not easily diffuse through it. That is why all commercial herbicides, etc., are less than 700 mol wt.

6.5.3 Plasma Membrane

Plasma membrane is the next barrier after the cell wall. Molecules can cross the plasma membrane by two processes:

i. Non-facilitated Diffusion – This process is based on physicochemical characteristics. It can be by a neutral lipophilic (lacking charge molecules), passive diffusion, down concentration gradient or lipophilic molecules with functional group sensitive to pH; weak acids can dissociate in less lipophilic ions and undissociated acids can freely move through plasma membrane.

ii. Carrier Mediated Diffusion – Cell membranes have receptor protein carriers, which are selective in the uptake of molecules, thus they limit the rate of uptake. They selectively carry out carrier-mediated transport through the membrane.

6.6 Carbon Dots as Trojan Horse to Penetrate Foliar Surface for Therapeutic Molecule Delivery

Though various nanomaterials and nanotechnologies have been applied in the last decade, comparatively few studies have focused on the effects, mechanisms and use of carbon dots (CDs) on plants. Although there has been enough information on the toxicity of CDs to animal and bacterial systems, only limited studies are available on higher plants. Higher plants strongly interact with the physicochemical factors of the environment, hence are also expected to be affected by the nano-sized particles. However, as mentioned in the previous chapter, studies on the toxicity of CDs have evidenced several positive effects on the growth and development of plantlets which have considered studies on seed germination and root elongation, often accompanied by other evaluations on biomass changes and anatomical-histological studies. But it is interesting to note that the results of Meziani *et al.* [402] have suggested that CDs can be activated by visible light (or even under ambient lighting conditions in rooms) for bactericidal activity.

Any particulate matter present in the atmosphere that gets deposited on the leaves causes an alteration in the transpiration rates, thermal balance and photosynthesis.

For foliar applied molecules, penetration and uptake can occur through cracks in the leaf surface, stomatal cavities, or by diffusion through the cuticle. Other surface features of the leaf, such as trichomes (hairs and glands), are also expected to greatly influence CD conjugated therapeutic molecules. Plants have trichomes on the leaf surface that can reduce CD uptake by intercepting spray droplets before they reach the epidermal surface of the leaf. The addition of a surfactant to the spray solutions allows drops to break into smaller droplets when they hit the trichomes, increasing the amount of spray solution contacting the epidermal surface. Using leaves of *Byrsonima sericea* and *Psidium guineense*, Da Silva *et al.* [403] have shown that these leaves have trichomes on the cuticular surface, which formed a barrier reducing the penetration of metal ions into the mesophyll, as observed by the lower iron leaf content and iron accumulation in trichomes. Whereas, based on the fact that penetration rates of foliar-applied polar solutes are highly variable (though the mechanism is not yet fully understood), Eichert *et al.* [404], using *Allium porrum* and *Vicia faba,* investigated the size exclusion limits and lateral heterogeneity of the stomatal foliar uptake pathway for aqueous solutes and water-suspended nanoparticles. The results suggested that the stomatal pathway fundamentally differs from the cuticular foliar uptake pathway.

Since CD is a water-soluble particle it can be envisaged as a nano-sized particle that can be used as a molecule delivery vehicle.

6.6.1 CD for Delivery of Antibiotics to Plants

A novel microwave-assisted synthesis of CDs from gum arabic (GA) by microwave-assisted method, their density gradient centrifugation (DGC) separation and finally their use as molecular vehicle to ferry *streptomycin,* a broad-spectrum antibiotic, has been reported by Sharon's group [72]. Due to exceptional biocompatibility, CDs are exploited as versatile drug delivery vehicles for chemotherapeutic payloads. Streptomycin is an amino-glycoside antibiotic, which binds irreversibly to bacterial ribosome (small 16S rRNA of 30S subunit of ribosome) and inhibits the protein synthesis. Streptomycin inhibits the growth of Gram-positive and Gram-negative bacteria. It also kills fungi and algae. It is used as an antibiotic for both animals and plants. Attachment of streptomycin to CD was achieved by stirring them both together at room temperature for 3 h at 30 × g. Attachment was confirmed by studying the change in the optical properties of Streptomycin @CD conjugate using UV-Vis spectroscopy in the spectra window of 200–600 nm with respect to pure CDs and using Fourier transform infrared (FTIR) and thermo-gravimetric analysis (TGA). Drug loading efficiency (DLE) of C-D was 75%. Based on data obtained following chemical bonds between CD and Streptomycin, conjugate is proposed to be as follows: either with the hydroxyl group or with the carboxyl group of the streptomycin (Figure 6.14):

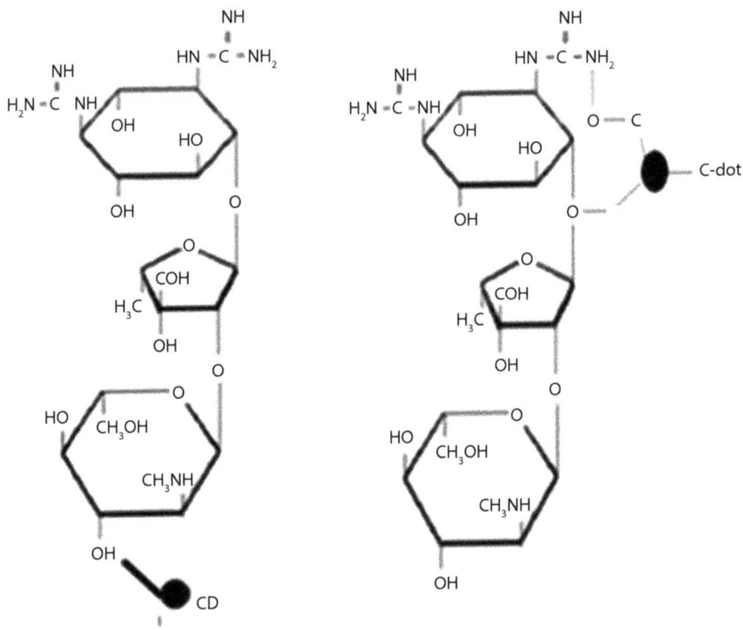

Figure 6.14 Possible proposed sites of attachment of streptomycin to carbon dot.

Streptomycin is known to control fire blight disease in apples, pears, etc; a disease that kills the plant. Foliar application of CD-Streptomycin is being tried. As we know, foliar application needs water-soluble materials so that it can be transported in the plants. Another advantage of foliar application is that it requires smaller quantities than when applied to the soil and the drugs applied to the foliage are absorbed more rapidly than when applied to the soil. However, the use of a high amount of antibiotics in agriculture is often not preferred because it is likely to develop other resistant phytopathogenic bacteria or develop resistant bacterial pathogenic strains of human pathogens. Apart from streptomycin, other chemicals that are used for treatment of fire blight are Cu-compounds, Mancozeb+Cu, Phosetyl Alluminium, Ca-Prohexadione, etc. At higher concentrations Cu is known to be phytotoxic.

CDs as a potential drug carrier to treat fire blight of apple – Fire blight of apple is a bacterial disease caused by *Erwinia amylovora*. This Gram-negative bacterium belongs to the same family, which includes *Escherichia coli, Yersinia* spp., *Shigella* spp. and *Salmonella* spp., which are human and animal pathogens. It is a rod-shaped bacterium of about 0.3 μm × 1–3 μm size (Figure 6.15). Apart from apples and pears, *E. amylovora* infects approximately 75 plant species of the *Rosaceae* family.

It enters the tree through the stomata, lenticels and hydathodes. It is a serious concern for growers of apples and pears.

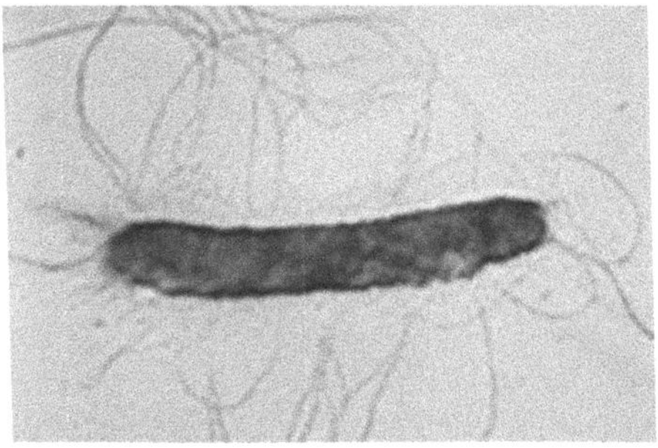

Figure 6.15 A rod-shaped *Erwinia amylovora* bacterium.

Once it infects the tree it travels through the phloem of the stems to the branches, leaves and apex, causing dieback. *E. amylovora* enters into the xylem or cortical parenchyma. If bacteria reach the *phloem*, they are carried upward to the tip of the twig and to the leaves. Invasion of a bacterial pore-forming complex becomes *lodged* in the host cell membrane.

The antibiotics used to treat fire blight are streptomycin, which is sprayed on the plants, or oxytetracycline, which is injected into them. The widespread use of streptomycin spray has led to antibiotic resistance.

A *fog spray* of CD conjugated with streptomycin is applied to the foliage of the infected apple tree during spring to mature trees and young plants in the nursery. Spraying can be done on the infected stem and branches in winter to save the tree, which is a proposed future activity of the authors of this book.

6.7 Concluding Remarks

This chapter has dealt with the possibility of using CD as a payload for nutrients/fertilizers and therapeutic agents, especially through foliar application. This approach will be very useful for microbes lodged in the phloem or vascular system of plants.

I would like to end this book with the following quote:

> *What we know is a drop, what we don't know is an ocean.*
>
> **Isaak Nyuton**

References

1. Planck M, *Ann. Phys.* 4, 553, 1901.
2. Bohr N., *Philos. Mag.* 26, 1, 1913.
3. Bohr N., *Philos. Mag.* 26, 476, 1913.
4. Schrödinger E, *Phys. Rev.* 28(6), 1049, 1926.
5. Kroto HW, Heath JR, O'Brien SCI., Curl RF, Smalley RE, *Nature* 318, 162, 1985.
6. Iijima S, *J. Cryst. Growth* 50, 675, 1980.
7. Iijima S, *Nature* 354, 56, 1991.
8. Geim AK., Novoselov KS, *Nat. Mater.* 6, 183, 2007.
9. Loh SM, Huang YH, Lin KM, Su WS, Wu BR, Leung TC, *Phys. Rev. B* 90, 035450, 2014.
10. Bhattacharya A, Chatterji S, Prajapati R, Mukherjee TK, *Phys. Chem. Chem. Phys.* 17, 12833, 2015.
11. Wu XL, Fan JY, Qui T, Yang X, Siu GG, Chu PK, *Phys. Rev. Lett.* 94, 026102, 2005.
12. Yeh TF, Huang LW, Chung CJ, Chiang IT, Teng H, *J. Phys. Chem. Lett.* 7, 2087, 2016.

13. Reed MA, Randall JN, Aggarwal RJ, Matyi RJ, Moore TM, Wetsel AE, *Phys. Rev. Lett.* 60(6), 535, 1988.
14. Alivisatos P, *Nat. Biotech.* 22(1), 47, 2004.
15. Smith AM, Duan WH, Mohs AM, Nie SM, *Adv. Drug Del. Rev.* 60(11), 1226, 2008.
16. Xing Y, Chaudry Q, Shen C, Kong KY, Zhau HE, Chung LW, Petros JA, O'Regan RM, Yezhelyev MV, Simons JW, Wang MD, Nie S, *Nat. Protoc.* 2(5), 1152–115, 2007.
17. Michalet X, Pinaud FF, Bentolila LA, Tsay JM, Doose S, Li JJ, Sundaresan G, Wu AM, Gambhir SS, Weiss S, *Science* 307(5709), 538–544, 2005.
18. Takeda M, Tada H, Higuchi HM, Kobayashi Y, Kobayashi M, Sakurai Y, Ishida T, Ohuchi N, *Breast Cancer* 15(2), 145, 2008.
19. Quarta A, Corato RD, Manna L, Ragusa A, Pellegrino T, *IEEE Trans. Nanobiosci.* 6(4), 298, 2007.
20. Dabbousi BO, Rodriguez-Viejo J, Mikulec FV, Heine JR, Mattoussi H, Ober R, Jensen KF, Bawendi MG, *J. Phys. Chem. B* 101(46), 9463, 1997.
21. Jin S, Hu Y, Gu Z, Liu L, Wu HC, *J. Nanomater.* 2011, 1, 2011.
22. Lin P, Chen JW, Chang LW, Wu JP, Redding L, Chang H, Yeh TK, Yang CS, Tsai MH, Wang HJ, Kuo YC, Yang RS, *Environ. Sci. Technol.* 42, 6264, 2008.
23. Larson DR, Zipfel WR, Williams RM, Clark SW, Bruchez MP, Wise FW, Webb WW, *Science* 300, 1434, 2003.
24. Cao L, Wang X, Meziani MJ, Lu FS, Wang HF, Luo PG, Lin Y, *J. Am. Chem. Soc.* 129, 11318–11319, 2007.
25. Xu X, Ray R, Gu Y, Ploehn HJ, Gearheart L, Raker K, Scirivens W, *J. Am. Chem. Soc.* 126, 12736, 2004.
26. Li H, Kang Z, Liu Y, Lee ST, *J. Mater. Chem.* 22, 24230, 2012.
27. Ponomarenko LA, Schedin F, Katnelson MI, Yang R, Hill EW, Novoselov KS, Geim AK, *Science* 320, 356, 2008.
28. Hsu PC, Chang HT, *Chem. Commun.*48, 3984–3986, 2012.
29. Li HT, Liu RH, Liu Y, Huang H, Yu H, Ming H, Lian SY, Lee ST, Kang ZH, *J. Mater. Chem.* 22, 17470, 2012.
30. Lim SY, Shen W, Gao Z, *Chem. Soc. Rev.* 44, 362, 2015.
31. Sharon M, Sharon M, *Graphene Science and Industry*, Wiley Scrivener, USA, 2015.

32. Zhou J, Booker C, Li R, Zhou X, Sham T, Sun X, Ding Z, *J. Am. Chem. Soc.* 129(4), 744, 2007.
33. Zheng LY, Chi YW, Dong YQ, Lin JP, Wang BB, *Am. Chem. Soc.* 131, 4564–4565, 2009.
34. Bao L, Zhang ZL, Tian ZQ, Zhang L, Liu C, Lin Y, Qi B, *Adv. Mater.* 23(48), 5801, 2011.
35. Que Y, Xiao W, Fei X, Chen H, Huang L, Du SX, Gao HJ, *Adv. Mater.* 21, 2777, 2009.
36. Li J, Wang Y, Sun Y, Ding C, Lin Y, Sun W, Luo C, *RSC Adv.*7, 2315–2322, 2017. doi:10.1039/C6RA25627A.
37. Li H, Ming H, Liu Y, Yu H, He X, Huang H, Pan K, Kang Z, Lee ST, *New J. Chem.* 35, 2666–2670, 2011.
38. Ming H., Ma Z, Liu Y, Pan K, Yu H., Wang F, Kang Z, *Dalton Trans.* 41(31), 9526–9531, 2012.
39. Yao S, Hu YF, Li GK, *Carbon* 66, 77, 2014.
40. Liu H, Ye T, Mao C, *Angew. Chem. Int. Ed.* 46(34), 6473, 2007.
41. Tian L, Ghosh D, Chen W, Pradhan S, Chang X, Chen S, *Chem. Mater.* 21(13), 2803–2809, 2009.
42. Ray SCI, Saha A, Jana NR, Sarkar R, *J. Phys. Chem. C* 113(43), 18546, 2009.
43. Bourlinos AB, Stassinopoulos A, Anglos D, Zboril R, Karakassides M, Giannelis EP, *Small* 4(4), 455, 2008.
44. Chen W, Hu C, Yang Y, Cui J, Liu Y, *Materials* 9, 184–191, 2016.
45. Hsu PC, Chang H-T, *Chem. Commun.* 48, 3984, 2012. http://dx.doi.org/ 10.1039/c2cc30188a.
46. Zhang B, Liu C, Liu Y, *Eur. J. Inorg. Chem.* 2010(28), 4411, 2010.
47. He X, Li H, Liu Y, Huang H, Kang Z, Lee S-T, *Colloids Surf. B* 87(2), 326, 2011.
48. Zhang YY, Wu M, Wang YQ, He XW, Li WY, Feng XZ, *Talanta* 117, 196–202, 2013.
49. Wei J, Qiang L, Ren J, Ren X, Meng X, Tang F, *Anal. Methods* 6, 1922, 2014.
50. Mewada A, Pandey S, Thakur M, Jadhav D, Sharon M, *J. Mater. Chem. B* 2, 698, 2014.
51. Zhang Z, Hao J, Zhang J, Zhang B, Tang J, *RSC. Adv.* 2, 8599, 2012.
52. Bhunia SK, Maity AR, Nandi S, Stepensky D, Jelinek R, *ChemBioChem.* 17, 614–619, 2016.

53. Peng J, Gao W, Gupta BK, Liu Z, Romero-Aburto R, Ge L, Li S, Alemany LB, Zhan X, Gao G, Vithayathil SA, Kaipparettu BA, Marti AA, Hayashi T, Zhu UJ, Pulickel MA, *Nano. Lett.* 12, 844–849, 2012.
54. Lee M, Lee J, Park SY, Min B, Kim B, In I, *Sci. Rep.* 5, 11707, 2015. doi:10.1038/srep11708.
55. Dong W, Dong Y, Wang Y, Zhou S, Ge X, Sui L, Wang J, *Spectrochim. Acta A Mol. Biomol. Spectrosc.*116, 209, 2013.
56. Sun YP, Zhou B, Lin Y, Wang W, Fernando KA, Pathak P, Meziani MJ, Harruff BA, Wang X, Wang H, Luo PJ, Yang H, Kose ME, Chen B, Veca LM, Xie SY, *J. Am. Chem. Soc.* 128, 7756, 2006.
57. Hu SL, Liu J, Yang JL, Wang YZ, Cao S, *J. Nanopart. Res.* 13, 7247, 2011.
58. Gonçalves H., Esteves da Silva JCG., *J. Fluoresc.* 20(5), 1023, 2010.
59. Habiba K, Marakov VI, Avlos J, Guniel MJF, Weiner BR, Morell G, *Carbon* 64, 341–350, 2013.
60. Bottini M, Balasubramanian C, Dawson MI, Bergamaschi A, Bellucci M, Mustelin T, *J. Phys. Chem. B* 110, 831, 2006.
61. Kargbo O, Jin Y, Ding SN, *Current Anal. Chem.* 11, 4, 2015.
62. Jiang H, Chen F, Lagally MG, Denes FS, *Langmuir* 26, 1991–1995, 2010.
63. Loh YH, Hartung O, Li H, Guo C, Sahalie JM, Manos PD, Urbach A, Heffner GC, Grskovic M, Vigneault F, Lensch MW, Park IH, Agarwal S, Church GM, Collins JJ, Irion S, Daley GQ, *Cell Stem Cell* 7(1), 15–19, 2010.
64. Li HT, He XD, Liu Y, Huang H, Lian S, Lee ST, Kang Z, *Carbon* 49, 605–609, 2011.
65. Li H, He X, Liu Y, Yu H, Kang Z, Lee ST, *Mater. Res. Bull.* 46(11), 147, 2011.
66. Pandey S, Mewada A, Oza G, Thakur M, Mishra N, Sharon M, Sharon M, *Nanosci. Nanotech. Lett.* 5, 1, 2013.
67. Zhu W, Wang X, Li Y, Wang Z, Yang F, Yang X, *Chem. Commun.* 2009, 5118–5120, 2009.
68. Qu S, Wang X, Lu Q, Liu X, Wang L, *Angew. Chem. Int. Ed.*124(49), 12381, 2012.

69. Du F, Zeng F, Ming Y, Wu S, *Microchim. Acta* 180(5–6), 453, 2013.
70. Xu M, He G, Li Z, He F, Gao F, Su Y, Zhang Z, Yang Z, Zhang Y, *Nanoscale* 6(17), 10307–10315, 2014. doi:10.1039/c4nr02792b.
71. Pandey S, Mewada A, Thakur M, Tank A, Sharon M, *RSC Adv.* 3, 26290, 2013.
72. Thakur M, Pandey S, Mewada S, Patil V, Khade M, Goshi E, Sharon M, *J. Drug Del.* 2014, 282193, 2014. http://dx.doi.org/10.1155/2014/282193.
73. Wang F, Xie Z, Zhang H, Liu CY, Zhang YG, *Adv. Funct. Mater.* 21, 1027–1031, 2011.
74. Wang X, Qu K, Xu B, Ren J, Qu X, *J. Mater. Chem.* 21(8), 2445, 2011.
75. Kwon W, Rhee SW, *Chem. Commun.* 48, 5256, 2012.
76. Zong J, Zhu YH, Yang XL, Shen JH, Li C, *Chem. Commun.* 47, 764, 2011.
77. Liu R, Wu D, Liu S, Koynov K, Knoll W, Li Q, *Agnew. Chem.* 121, 4668–4671, 2009.
78. Bourlinos AB, Stassinopoulos A, Anglos D, Zboril R, Karakassides M, Giannelis EP, *Chem. Mater.* 20(14), 4539, 2008.
79. Himaja AL, Karthik PS, Sreedhar B, Singh SP, *J. Fluoresc.* 24(6), 1767–1773, 2014.
80. Sahu S, Behera B, Maiti TK, Mohapatra S, *Chem. Commun.* 48(70), 8835–8837, 2012.
81. Hermann R, Schwarz H, Müller M, *J. Struct. Biol.* 107, 38–47, 1991.
82. Guan W, Gu W, Ye L, Guo C, Su S, Xu P, Xue M, *Int. J. Nanomed.* 9, 5071–5078, 2014.
83. Phadke C, Mewada A, Dharmatti R, Thakur M, Pandey S, Sharon M, *J. Flouresc.* 25(4), 1103–1107, 2015.
84. Wang X, Cao L, Lu F, Meziani M, Li H, Qi G, Zhou B, Sun YP, *ChemComm* 25, 3774–3776, 2009.
85. Roshni V, Ottur D, *Curr. Science* 112(2), 385–390, 2017.
86. Mao XJ, Zheng HZ, Long YJ, Du J, Hao JY, Wang LL, Zhou DB, *Spectrochim. Acta Part A* 75, 553–557, 2010.
87. Platt G, Damin F, Swann M, Metton I, Skorski G, Cretich M, Chiari M, *Biosens. Bioelectron.* 52, 82–88, 2014.

88. Zheng H, Wang Q, Long Y, Zhang H, Huang X, Zhu R, *Chem. Commun.* 47, 10650–10652, 2011.
89. Bourlinos AB, Stassinopoulos A, Anglos D, Zboril R, Georgakilas V, Giannelis EP, *Chem. Mater.* 20(14), 4539–4541, 2008.
90. Dong Y, Shao J, Chen C, Li H, Wang R, Chi Y, Lin X, Chen G, *Carbon* 50, 4738–4743, 2012.
91. Kim S, Hwang SW, Kim MK, Shin DY, Shin DH, Kim CO, Yang SB, Park JH, Hwang E, Choi SH, Ko G, Sim S, Sone C, Choi HJ, Bae S, Hong BH, *ACS Nano* 6(9), 8203–8208, 2012. http://dx.doi.org/10.1021/nn302878r.
92. Peng H, Travas-Sejdic J, *Chem. Mater.* 21, 5563, 2009.
93. Li HL, Zhang YW, Wang L, Tian JQ, Sun XP, *Chem. Commun.* 47, 961–963, 2011.
94. Yang ZC, Wang M, Yong AM, Wong SY, Zhang XH, Tan H, Chang AY, Li X, Wang J, *Chem. Commun.* 47, 11615, 2011.
95. Tang L, Ji R, Cao X, Lin J, Jiang H, Li X, Teng KS, Luk CM, Zeng S, Hao J, Lau SP, *ACS Nano* 6(6), 5102–5110, 2012. http://dx.doi.org/10.1021/nn300760g.
96. Bourlinos AB, Bakandritsos A, Kouloumpis A, Gournis D, Krysmann M, Giannelis EP, Polakova K, Safarova K, Hola K, Zboril R, *J. Mater. Chem.* 22, 23327–23330, 2012.
97. Puvvada N, Prashanth-Kumar BN, Konar S, Kalita H, Mandal M, Pathak A, *Sci. Technol. Adv. Mater.* 13(4), 045008, 2012.
98. Li H, Zhai J, Sun X, *Langmuir* 27, 4305, 2011.
99. Wang L, Wang X, Bhirde A, Cao J, Zeng Y, Huang X, Sun Y, Liu G, Chen X, *Adv. Healthcare Mater.* 3, 1203–1209, 2014.
100. Hu L, Sun Y, Li S, Wang X, Hu K, Wang L, Liang XJ, Wu Y, *Carbon* 67, 508–513, 2014.
101. Yang ST, Cao L, Luo PGJ, Lu F, Wang X, Wang H, Meziani MJ, Liu Y, Qi G, Sun YP, *J. Am. Chem. Soc.* 131, 11308–11309, 2009.
102. Qu L, Peng X, *J. Am. Chem. Soc.* 124(9), 2049–2055, 2002.
103. Talapin D, Rogach AL, Kornowski A, Haase M, Weller H, *Nano Lett.* 4(1), 2017–211, 2001.
104. Bruchez M Jr, Moronne M, Gin P, Weiss S, Alivisatos AP, *Science* 281(5385), 2013–2016, 1998.
105. Gu L, Hall DJ, Qin Z, Anglin E, Joo J, Mooney DJ, Howell SB,Sailor MJ, *Nat. Commun.* 4, 2326, 2013.

106. Thakur M, Mewada A, Pandey S, Bhori M, Singh K, Sharon M, Sharon M, *Mater. Sci. Eng. C* 67, 468–477, 2016.
107. Jia XF, Li J, Wang EK, N*anoscale* 4, 5572–5575, 2012.
108. Zhang H, Yang, Hou H, Chen S, Yao H, *Sci. Rep.* 7, 7769, 2017. doi:10.1038/s41598-017-07345-y
109. Li HT, Liu RH, Liu Y, Huang H, Yu H, Ming H, Lian SY, Lee ST, Kang ZH, *J. Mater. Chem.* 22, 17470, 2012.
110. Fukui K, Yonezawa T, Shingu H, *J. Chem. Phys.* 20, 722, 1952.
111. Joaquim C.G., da Silva E, Golcalves HMR, *Trends Anal. Chem.* 30(8), 1327, 2011.
112. Demchenko AP, Dekaliuk MO, *Methods Appl. Fluoresc.* 1, 042001, 2013.
113. Wang W, Damm C, Walter J, Nacken TJ, Peukert W, *Phys. Chem. Chem. Phys.* 18, 466–475, 1, 2016.
114. Khan S, Verma NC, Gupta A, Nandi CK, *Sci. Rep.*5, 11423, 2015.
115. Li Q, Ohulchanskyy TY, Liu R, Koynov K, Wu D, Best A, Kumar R, Bonoiu A, Prasad PN, *J. Phys. Chem. C* 114(28), 12062–12068, 2010.
116. Salinas-Castillo A, Ariza-Avidad M, Pritz C, Camprubí-Robles M, Fernández B, Ruedas-Rama MJ, *et al.*, *Chem. Commun.* 49(11), 1103–1105, 2013.
117. Li JF, Huang JF, Ding Y, Yang ZL, Li SB, Zhou XS, Fan FR, Zhang W, Zhou ZY, Wu DY, Ren B, Wang ZL, Tian ZQ, *Nature* 464, 392–395, 2010.
118. Li H, He X, Kang Z, Hui H, Liu Y, Liu J, Lian S, Tsang CHA, Yang X, Lee ST, *Angew. Chem. Int. Ed.* 49, 4430–4434, 2010.
119. Chen XB, Mao SS, *Chem. Rev.* 107, 2891, 2007.
120. Li JY, Ma WH, Chen CC, Zhao JC, Zhu HY, Gao XP, *J. Mol. Catal. A: Chem.* 261, 131, 2007.
121. Mehta A, Nelson EJ, Webb SM, Holt JK, *Adv. Mater.* 21, 102, 2009.
122. Zhang HC, Ming H, Lian S, Huang H, Li H, Zhang L, Liu Y, Kang Z, Lee ST, *Dalton Trans.* 40, 10822, 2011.
123. Laraoui A, Hodges JS, Meriles CA, *Nano Lett.* 12, 3477–3482, 2012.
124. Shen P, Xia Y, *Anal. Chem.* 86, 5323–5329, 2014.
125. Xue W, Lin Z, Chen H, Lu C, Lin JM, *J. Phys. Chem. C* 115, 21707–21714, 2011.

126. Borjanović V, Bistričić L, Pucić I, Mikac L, Slunjski R, Jakšić M, McGuire G, Stanković AT, Shenderova O, *J. Mater Sci.* 51, 1000–1016, 2016.
127. Hola K, Bourlinos AB, Kozak O, Berka K, Siskova KM, Havrdova M, Tucek J, Safarova K, Otyepka M, Giannelis EP, Zboril R, *Carbon* 70, 279–286, 2014.
128. Zhao D, Zhao CQ, Li M, Ren JS, Qu XG, *Anal. Chim. Acta* 809, 128–133, 2014.
129. Zhang YL, Wang L, Zhang HC, Liu Y, Wang HY, Kang ZH, Lee ST, *RSC Adv.* 3, 3733–3738, 2013.
130. Mewada A, Pandey S, Shinde S, Mishra N, Oza G, Thakur M, Sharon M, Sharon M, *Mater. Sci. Eng. C.* 33, 2914–2917, 2013.
131. Pandey S, Thakur M, Mewada A, Anjarlekar D, Mishra N, Sharon M, *J. Mater. Chem. B* 1, 4972–4982, 2013.
132. Chan WCW, Nie S, *Science* 281, 2016–2018, 1998.
133. Han X, Jing Z, Wu W, Zou B, Peng X, Ren P, Wikramanayake A, Lu Z, Leblanc RM, *Nanoscale* 9, 12862–12866, 2017.
134. Kroll RA, Pagel MA, Muldoon LL, Roman-Goldstein S, Fiamengo SA, Neuwelt EA, *Neurosurgery* 43, 879–886, 1998.
135. Redhead HM, Davis SS, Illum L, *J. Control. Release* 70, 353–363, 2001.
136. Chowdhury D, Gogoi N, Majumdar G, *RSC Adv.* 2, 12156–12159, 2012.
137. Yang ST, Wang X, Wang HF, Lu F, Luo PG, Cao L, Meziani MJ, Liu J-H, Liu Y, Chen M, Huang Y, Sun YP, *J. Phys. Chem. C* 113, 18110–18114, 2009.
138. Lee DY, Khatun Z, Lee JH, Lee YK, *Biomacromolecules* 2, 336–341, 2011.
139. Chandra S, Patra P, Pathan SH, Roy S, Mitra S, Layek A, Bhar R, Pramanik P, Goswami A, *J. Mater. Chem. B* 1, 2375–2382, 2013.
140. Diac A, Focsan M, Socaci C, Gabudean AM, Farcau C, Maniu D, Vasile E, Terec A, Veca LM, Astilean S, *RSC Adv.* 5(95), 77662–77669, 2015.
141. Saha K, Bajaj A, Duncan B, Rotello VM, *Small* 7, 1903–1918, 2011.
142. Elghanian R, Storhoff JJ, Mucic RC, Letsinger RL, Mirkin CA, *Science* 277, 1078–1081, 1997.

143. De M, Rana S, Akpinar H, Miranda OR, Arvizo RR, Bunz UH, Rotello VM, *Nat. Chem.* 1, 461–465, 2009.
144. Dubertret B, Skourides P, Norris D J, Noireaux V, Brivanlou AH, Libchaber A, *Science* 298, 1759–1762, 2002.
145. Wu X, Liu H, Liu J, Haley KN, Treadway JA, Larson JP, Ge N, Peale F, Bruchez MP, *Nat. Biotechnol.*21, 41–46, 2003.
146. Lagerholm BC, Wang MM, Ernst LA, Ly DH, Liu HJ, Bruchez MP, Waggoner AS, *Nano Lett.* 4(10), 2019–2022, 2004.
147. Lu Q, Moore JM, Huang G, Mount AS, Larcom L, Ke PC, *Nano Lett.* 4, 2473–2477, 2004.
148. Santra S, Yang H, Stanley JT, Holloway PH, Moudgil BM, Walter G, Mericle RA, *Chem. Commun.* 25, 3144–3146, 2005.
149. Stroh M, Zimmer JP, Duda DG, Levchenko TS, Cohen KS, Brown EB, Scadden DT, Torchilin VP, Bawendi MG, Fukumura D, Jain RK, *Nat. Med.* 11, 678–682, 2005.
150. Zheng XT, Than A, Ananthanaraya A, Kim DH, Chen P, ACS Nano, 7: 6278, 2013.
151. Hu Q, Man CP, Zhang Y, Wan C, Gong X, Zhang L, Choi MMF, *J. Chromatogr. A* 1304, 234–240, 2013.
152. Milosavljevic V, Nguyen HV, Michalek P, Moulick A, Kopel P, Kizek R, Adam V, *Chem. Pap.* 69(1), 192–201, 2015.
153. Ding H, Du F, Liu P, Chen Z, Shen J, *ACS Appl. Mater. Interfaces* 7(12), 6889–6897, 2015.
154. Singh S, Mishra A, Kumari R, Sinha KK, Singh MK, Das P, *Carbon*, 114: 169–176, 2017.
155. Xu B, Zhao C, Wei W, Ren J, Miyoshi D, Sugimoto N, Qu X, *Analyst* 137(23), 5483–5486, 2012.
156. Kim J, Park J, Kim H, Singha K, Kim WJ, *Biomaterials* 34, 7168e-7180, 2013.
157. Liu H, Wang Q, Shen G, Zhang C, Li C, Ji W, Wang C, Cui D, *Nanoscale Res. Lett.* 9, 397–406, 2014.
158. Wang Q, Zhang C, Shen G, Liu H, Fu H, Cui D, *J. Nanobiotechnol.* 12, 58, 2014.
159. Wu Y, Wei P, Pengpukiat S, Schumacher EA, Remcho VT, *Anal. Chem.* 87, 8510, 2015.
160. Nandi S, Malishev R, Bhunia SK, Kolusheva S, Jopp J, Jelinek R, *Biophysical J.* 110, 2016–2025, 2016.
161. Nandi S, Ritenberg M, Jelinek R, *Analyst* 140, 4232, 2015.

162. Campos BB, Oliva MM, Contreras-Cáceres R, Rodriguez-Castellón E, Jiménez-Jiménez J, Esteves da Silva JCG, Algarra M, *J. Colloid Interface Sci.* 458, 209–218, 2015.
163. Li CL, Ou CM, Huang CC, Wu WC, Chen YP, Lin TE, Ho LC, Wang CW, Shih CC, Zhou HC, Lee YC, Tzeng WF, Chiou TJ, Chu ST, Cang J, Chang HT, *J. Mater. Chem. B* 2, 4564, 2014. http://dx.doi.org/10.1039/C4TB00216D.
164. Zhou L, He B, Huang J, *Chem. Commun.* 49, 8078–8080, 2013.
165. Wang X, Cao L, Yang ST, Lu F, Meziani MJ, Tian L, Sun KW, Bloodgood MA, Sun YP, *Angew Chem. Int. Ed. Engl.* 49, 5310–5314, 2010.
166. Sachadev A, Matai I, Kumar SU, Bhushan B, Dubey P, Gopinath P, *RSC Adv.* 3, 16958–16961, 2013.
167. Cayuela A, Kennedy SR, Soriano ML, Jones CD, Valcárcel M, Steed JW, *Chem. Sci.* 6, 6139–6146, 2015.
168. Gogoi N, Barooah M, Majumdar G, Chowdhury D, *ACS Appl. Mater. Interfaces* 7, 3058–3067, 2015.
169. Wang JL, Su SH, Wei JH, Bahgi R, Hope-Weeks L, Qiu JJ, *Physica E* 72, 17–24, 2015.
170. Konwar A, Gogoi N, Majumdar G, Chowdhury D, *Carbohydr. Polym.* 115, 238–245, 2015.
171. Campos BB, Mutavdžić D, Stanković M, Radotić K, Lázaro-Martínez JM, Esteves da Silva JCG, Cáceres RC, González MSP, Castellóne ER, Algarra M, *New J. Chem.* 41, 4835–4842, 2017.
172. Krishna AS, Radhakumary C, Antony M, Sreenivasan K, *J. Mater. Chem. B* 2, 8626–8632, 2014.
173. Krishna AS, Radhakumary C, Sreenivasan K, *Anal. Methods* 7, 9482–9488, 2015.
174. Sun YP, Wang X, Lu F, Cao L, Meziani MJ, Luo PG, Gu L, Veca M, *J. Phys. Chem. C* 112, 18295–18298, 2008.
175. Cao L, Sahu S, Anikumar P, Bunker CE, Xu J, Shiral Fernando K, Wang P, Guliants EA, Tackett KN, Sun YP, *J. Am. Chem. Soc.* 133, 4754–4757, 2011.
176. Gupta V, Chaudhary N, Srivastava R, Sharma GD, Bhardwaj R, Chand S, *J. Am. Chem. Soc.* 133, 9960–9963, 2011.
177. Xu Q, Kuang T, Liu Y, Cai L, Peng X, Sreeprasad TS, Zhao P, Yu Z, Li N, *J. Mater. Chem. B* 4, 7204–7219, 2016.

178. Kong B, Zhu AW, Ding CQ, Zhao XM, Li B, Tian Y, *Adv. Mater.* 24, 5844–5848, 2012.
179. Luo P, Li C, Shi G, *Phys. Chem. Chem. Phys.* 14, 7360–7366, 2012.
180. Dharmatti R, Phadke C, Mewada A, Thakur M, Pandey S, Sharon M, *Mater. Sci. Eng. C* 44, 9298, 2014.
181. Pandey S, Oza G, Mewada A, Shah R, Thakur M, Sharon M, *J. Mater. Chem. B*1, 1361–1370, 2013.
182. Lai CW, Hsiao YH, Peng YK, Chou PT, *J. Mater. Chem.* 22, 14403, 2012.
183. Han X, Han Y, Hui H, Zhang H, Zhang X, Liu R, Liu Y, Kang Z, *Dalton Trans.* 49, 10380–10383, 2013.
184. Pandey S, Mewada A, Thakur M, Pillai S, Dharmatti R, Sharon M, *RSC Adv.* 4, 1174, 2014.
185. Li J, Zhang B, Wang F, Liu C, *New J. Chem.* 35, 554–557, 2011.
186. Li C, Zhu Y, Zhang X, Yang X, Li C, *RSC Adv.* 2, 1765–1768, 2012.
187. Li BJ, Cao HQ, *J. Mater. Chem.* 21, 3346–3349, 2011.
188. Liu C, Zhang P, Zhai X, Tian F, Li W, Yang J, Liu Y, Wang H, Wang W, Liu W, *Biomaterials* 33(13), 3604–3613, 2012.
189. Yin JY, Liu HJ, Jiang S, Chen Y, Yao Y, *ACS Macro Lett.* 2, 1033–1037, 2013.
190. Liu Y, Yu YX, Zhang WD, *J. Alloys Compd.* 569, 102–110, 2013.
191. Yu BY, Kwak SY, *J. Mater. Chem.* 22, 8345–8353, 2012.
192. Wang F, Liu Y, Ma Z, Li HT, Kang ZH, Shen MR, *New J. Chem.* 37, 290–294, 2013.
193. Fu CC, Hung TC, Chen JY, Su CH, Wu WT, *Bioresour. Technol.* 101(22), 8750–8754, 2010.
194. Tangy A, Kumar VB, Pulidindi IN, Kinel-Tahan Y, Yehoshua Y, Gedanken A, *Energy Fuels* 30(12), 10602–10610, 2016.
195. Rehor I, Vilimova V, Jendelova P, Kubicek V, Jirak D, Herynek V, Kapcalova M, Kotek J, Cerny J, Hermann P, Lukes I, *J. Med. Chem.* 54, 5185–5194, 2011.
196. Li IF, Su CH, Sheu HS, Chiu HC, Lo HW, Lin WT, Chen JH, Yeh CS, *Adv. Funct. Mater.*18, 766–776, 2008.
197. Huh HS, Lee SW, *Bull. Korean Chem. Soc.* 27, 1839–1843, 2006.
198. Sastry RK, Rashmi HB, Rao NH, Ilyas SM, *Technol. Forecast. Soc. Change* 77, 639–648, 2010.

199. Zhao HX, Liu LQ, Liu ZD, Wang Y, Zhao XJ, Huang CZ, *Chem. Commun.* 47, 2604, 2011.
200. Song Y, Chen J, Hu D, Liu F, Li P, Li H, Chen S, Tan H, Wang L, *Sens. Actuators B Chem.* 221, 586–592, 2015.
201. D'souza SL, Deshmukh B, Bhamore JR, Rawat KA, Lenka N, Kailasa S.K, *RSC Adv.* 6, 12169, 2016.
202. Dong Y, Wang R, Li G, Chen C, Chi Y, Chen G, *Anal. Chem.* 84, 6220, 2012.
203. Huang XL, Zhang F, Zhu L, Choi KY, Guo N, Guo JX, Tackett K, Anilkumar P, Liu G, Quan QM, Choi HS, Niu G, Sun Y-P, Lee S, Chen XY, *ACS Nano* 7(7), 5684–5693, 2013.
204. Sun Y, Chen S, Wang J, Lu Y, *RSC Adv.* 5, 16368–16375, 2015. doi:10.1039/C4RA13820A.
205. Jana D, Sun CL, Chen LC, Chen KH, *Progr. Mater. Sci.* 58, 565, 2013.
206. Chandra S, Das P, Bag S, Laha D, Pramanik P, *Nanoscale* 3, 1533, 2011.
207. Zhou J, Shan X, Ma J, Gu Y, Qian Z, Chen J, Feng H, *RSC Adv.* 4, 5465, 2014.
208. Han D, Du F, Liu P, Chen Z, Shen J, *ACS Appl. Mater. Interfaces* 7(12), 6889–6897, 2015.
209. Chen H, Wang Z, Zong S, Chen P, Zhu D, Wu L, Cui Y, *Nanoscale* 7, 15477, 2015.
210. Liu JH, Anilkumar P, Cao L, Wang X, Yang ST, Luo PG, Wang H, Lu F, Meziani MJ, Liu Y, Korch K, Sun YP, *Nano Life* 1&2, 153–161, 2010.
211. Wang Y, Anilkumar P, Cao L, Liu JH, Luo PG, Tackett KN, Sahu S, Wang P, Wang X, Sun YP, *Exp. Biol. Med.* 236, 1231–1238, 2011.
212. Murali VS, Wang R, Mikoryak CA, Pantano P, Draper R, *Exp. Biol. Med.* 240(9), 1147–1151, 2015.
213. Zhao QL, Zhang ZL, Huang BH, Peng J, Zhang M, Pang DW, *Chem Commun.* 41, 5116–5118, 2008.
214. Wang YH, Wang SM, Ge SG, Wang SW, Yan M, Zang DJ, Yu JH, *Anal. Methods-UK* 5, 1328, 2013.
215. Yang Y, Wu D, Han S, Hu P, Liu R, *Chem. Commun.* 49, 4920–4922, 2013.
216. Boussif O, *Proc. Natl. Acad. Sci.* 92, 7297–7301, 1995.

217. Han B, Wang W, Wu H, Fang F, Wang N, Zhang X, Xu S, *Colloids Surf. B* 100, 209–214, 2012.
218. Dou Q, Fang X, Jiang S, Chee PL, Lee TC, Loh XJ, *RSC Adv.* 5, 46817–46822, 2015.
219. Wu YF, Wu HC, Kuan CH, Lin CJ, Wang LW, Chang CW, Wang TW, *Sci. Rep.* 6, 21170, 2016. doi:10.1038/srep21170.
220. Mondal S, Purkayastha P, *J. Phys. Chem. C* 120(26), 14365–14371, 2016.
221. Singer TP, Oxidative metabolism of cysteine and cystine, in: *Metabolic Pathways, 3rd ed., vol. 7, Metabolism of Sulfur Compounds*, Greenberg DM (ed.), Academic Press: New York, p. 545, 1975. ISBN 9780323162081.
222. Kuruvilla SJ, Li S, Sansalone L, Fortes B, Zheng I, Blackwelder P, Pumilia C, Micic M, Orbulescu J, Leblanc RM, *J. Parkinsons Dis.Alzheimers Dis.* 2(1), 7, 2015.
223. Lee CC, MacKay JA, Frechet JMJ, Szoka FC, *Nat. Biotechnol.* 23, 1517–1526, 2005.
224. Stiriba SE, Frey H, Haag R, *Angew. Chem. Int. Ed.* 41, 1329–1334, 2002.
225. Svenson S, Tomalia DA, *Adv. Drug Del. Rev.* 57, 2106–2129, 2005.
226. Kang KS, Lee HU, Kim MI, Park SY, Chang SJ, Park JH, Huh YS, Lee J, Yang M, Lee YC, Park HG, *J. Nanobiotechnol.* 13, 88, 2015.
227. Bai J, Sun C, Jiang X, *Anal. Bioanal. Chem.* 408(17), 4705–4714, 2016.
228. Yang J, Zhang X, Ma YH, Gao G, Chen X, Jia HR, Li YH, Chen Z, Wu FG, *ACS Appl. Mater. Interfaces* 8(47), 32170–32181, 2016.
229. Goodrum LJ, Patel A, Leykam JF, Kieliszewski MJ, *Phytochemistry* 54(1), 99–106, 2000.
230. Norrby SR, Nord CE, Finch R, *Lancet Infect. Dis.* 5(2), 115–119, 2005.
231. Gogoi N, Chowdhury D, *J. Mater. Chem. B* 2, 4089, 2014.
232. Zhong D, Zhuo Y, Feng Y, Yang X, *Biosens. Bioelectron.* 74, 546–555, 2015.
233. Jijie R, Barras A, Teodorescu F, Boukherroub R, Szunerits S, *Mol. Syst. Des. Eng.* 2, 349–369, 2017.

234. Gabathuler R, *Neurobiol. Dis.* 37, 48–57, 2010.
235. Upadhyay RK, *Int. Sch. Res. Notices* 2014, Article ID 309404, 2014. doi:10.1155/2014/309404.
236. Fletcher J, Huehns E, *Nature* 218, 1211–1214, 1968.
237. Qian ZM, Li H, Sun H, Ho K, *Pharmacol. Rev.* 54, 561–587, 2002.
238. Gan CW, Feng SS, *Biomaterials* 31, 7748–7757, 2010.
239. Anabousi S, Bakowsky U, Schneider M, Huwer H, Lehr CM, Ehrhardt C, *Eur. J. Pharm. Sci.* 29, 367–374, 2006.
240. Oh S, Kim BJ, Singh NP, Lai H, Sasaki T, *Cancer Lett.* 274, 33–39, 2009.
241. Xu Q, Liu Y, Su S, Li W, Chen C, Wu Y, *Biomaterials* 33, 1627–1639, 2012.
242. Li S, Amat D, Peng Z, Vanni S, Raskin S, De Angulo G, Othman AM, Graham RM, Leblanc RM, *Nanoscale* 8, 16662, 2016.
243. Li S, Peng Z, Dallman J, Baker J, Othman AM, Blackwelder PL, Leblanc RM, *Colloids Surf. B Biointerfaces* 145, 251–256, 2016.
244. Shi Y, Li C, Liu S, Liu Z, Zhu J, Yang J, Hu X, *RSC. Adv.* 5, 64790–64796, 2015.
245. Zhang Q, Zhang C, Li Z, Ge J, Li C, Dong C, Shuang S, *RSC. Adv.* 5, 95054–95060, 2015.
246. Lawaczeck R, Nandi PK, Nicolau C, *Biochim. Biophys. Acta* 903, 123–131, 1987.
247. Seksek O, Bolard J, *J. Cell Sci.* 109, 257–262, 1996.
248. Zimmermann K, Intracellular transport in cancer treatments: Carbon nanohorns conjugated to quantum dots and chemotherapeutic agents, MSc Thesis, Virginia Polytechnic Institute and State University, 2012.
249. Jung YK, Shin E, Kim BS, *Sci. Rep.* 5, 18807, 2015.
250. Yang L, Wang Z, Wang J, Jiang W, Jiang X, Bai Z, He Y, Jiang J, Wang D, Yang L, *Nanoscale* 8(12), 6801–6809, 2016b.
251. Zhang M, Yuan P, Zhou N, Su Y, Shao M, Cheng C, *RSC. Adv.* 7, 9347, 2017.
252. Yang XD, Xiang HJ, Yang XP, Liu JG, *New J. Chem.* 39, 800–804, 2015.
253. Zheng M, Liu S, Li J, Qu D, Zhao H, Guan X, Hu X, Xie Z, Jing X, Sun Z, *Adv. Mater.* 26(21), 3554–3560, 2014.

254. Jing YJ, Zhu YH, Yang XL, Shen JH, Li CZ, *Langmuir* 27, 1175–1180, 2011.
255. Tang J, Kong B, Wu H, Xu M, Wang YC, Wang YL, Zhao D, Zheng G, *Adv Mater.* 25(45), 6569–6574, 2013.
256. Karthik S, Saha B, Ghosh SK, Singh NDP, *Chem. Commun.* 49(89), 10471–10473, 2013.
257. Liu XJ, LK Pan, Zhao QF, Lv T, Zhu G, Chen TQ, Lu T, Sun ZH, Sun CQ, *Chem. Eng. J.* 183, 238, 2012.
258. Cheng L, Li Y, Zhai X, Xu B, Cao Z, Liu W, *ACS Appl. Mater. Interfaces* 6(22), 20487–20497, 2014.
259. Weinberg RA, *The Biology of Cancer*, Taylor and Francis Group, London, ISBN: 10 0-8153-4078-8, 2008.
260. Shankar E, Kanwal R, Candamo M, Gupta S, *Semin. Cancer Biol.* 40–41, 82–99, 2016.
261. Zhao A, Chen Z, Zhao C, Gao N, Ren J, *Carbon* 85, 309–327, 2015.
262. Wang W, Lu YC, Huang H, Feng JJ, Chen JR, Wang AJ, *Analyst* 139(7), 1692–1696, 2014. doi:10.1039/c3an02098c.
263. Wang, Q, Huang, X, Long Y, Wang X, Zhang H, Zhu R, Liang L, Teng P, Zheng H, *Carbon* 59, 192–199, 2013.
264. Wang Z, Liu G, Zheng H, Chen X, *Biotechnol. Adv.* 32(4), 831–843, 2013.
265. Bagul R, Mahajan V, Dhake A, *Int. J. Curr. Pharma. Res.* 4(3), 29–38, 2012.
266. Wang, H, Di J, Sun Y, Fu J, Wei Z, Matsui H, Alonso DC, Zhou A, *Adv. Funct. Mater.* 25, 5537–5547, 2015.
267. Cui X, Zhu L, Wu J, Hou Y, Wang P, Wang Z, Yang M, *Biosens. Bioelectron.* 63, 506–512, 2015.
268. Jain KK, *The Handbook of Nanomedicine*, Human Press, Springer, London, ISBN: 978-1-61779-983-9, 2012.
269. Palashuddin SK, Jaiswal A, Paul A, Ghosh SS, Chattopadhyay A, *Sci. Rep.* 2, 2012.
270. Peng H, Sejdic JT, *Chem. Mater.* 21(23), 5563–5565, 2009.
271. Qin X, Lu W, Asiri AM, Al-Youbi AO, Sun X, *Catal. Sci. Technol.* 3, 1027–1035, 2013.
272. Amjadi M, Mostaghaci B, Sitti M, *Curr. Gene Ther.* 17, 139–146, 2017.

273. Patra S, Roy E, Madhuri R, Sharma PK, *Biomater. Sci.*4(3), 418–429, 2016.
274. Guldi DM, Illescas BM, Atienza CM, Wielopolski M, Martin N, *Chem. Soc. Rev.* 38, 1587–1597, 2009.
275. De Volder MFL, Tawfick SH, Baughman RH, Hart AJ, *Science* 339, 535–539, 2013.
276. Zubair M, Mustafa M, Ali A, Doh YH, Choi KH, *J. Mater. Sci. Mater. Electron.* 26, 3344–3351, 2015. doi:10.1007/s10854-015-2837-2.
277. Wang J, Qiu J, *Sci. Adv. Mater.* 7, 1979–1989, 2015.
278. Wang X, Pei Y, Lu M, Lu X, Du XJ, *Mater Sci.* 50, 2113–2121, 2015.
279. Georgakilas V, Perman JA, Tucek J, Zboril R, *Chem Rev.* 115, 4744–4822, 2015.
280. Baker SN, Baker GA, *Angew. Chem. Int. Ed.* 49, 6726–6744, 2010.
281. Zhang ZP, Zhang J, Chen N, Qu LT, *Energy Environ. Sci.* 5, 8869–8890, 2012.
282. Wang Y, Anilkumar P, Cao L, Liu JH, Luo PG, Tackett KN, Sahu S, Wang P, Wang X, Sun YP, *Exp. Biol. Med.* 236, 1231–1238, 2011.
283. Xu JY, Zhou Y, Cheng GF, Dong MT, Liu SX, Huang CB, *Luminescence* 30, 411–415, 2015.
284. Kasibabu BSB, D'souza SL, Jha S, Singhal RK, Basu H, Kailasa SK, *Anal. Methods* 7, 2373–2378, 2015.
285. Jiang H, Chen F, Lagally MG, Denes FS, *Langmuir* 26, 1991–1995, 2009.
286. Qian Z, Ma J, Shan X, Feng H, Shao L, Chen J, *Chem. Eur. J.* 20(8), 2254–2263, 2014.
287. Nie S, *Nanomedicine* 5(4), 523–528, 2010.
288. Tao H, Yang K, Ma Z, Wan J, Zhang Y, Kang Z, Liu Z, *Small* 8(2), 281–290, 2012.
289. Nurunnabu M, Khatun Z, Huh KM, Park SY, Lee DY, Cho KJ, Lee Y, *ACS Nano* 7(8), 6858–6867, 2013.
290. Liu XJ, Guo ML, Huang J, Yin XY, *Bioresources* 8, 2537, 2013.
291. Deng Y, Chen X, Wang F, Zhang X, Zhao D, Shen D, *Nanoscale* 6, 10388, 2014.

292. Lu S, Guo S, Xu P, Li X, Zhao Y, Gu W, Xue M, *Int. J. Nanomedicine* 11, 6325–6336, 2016.
293. Xiao Y, Forry SP, Gao X, Holbrook RD, Telford WG, Tona A, *J. Nanobiotechnology* 8, 13, 2010.
294. Li H, Kang Z, Liu Y, Lee S, *J. Mater. Chem.* 22, 24230–24253, 2012.
295. Ding H, Cheng LW, Ma YY, Kong JL, Xiong HM, *New J. Chem.* 37(8), 2515–2510, 2013.
296. Hsu PC, Chen PC, Ou CM, Chang HY, Cang HT, *J. Mater. Chem. B* 1, 1774–1781, 2013.
297. Puvadda N, Prashanth-Kumar BN, Konar S, Kalita H, Mandal M, Pathak A, *Sci. Technol. Adv. Mater.*13(4), 045008, 2012.
298. Goh EJ, Kim KS, Kim YR, Jung HS, Beack S, Kong WH, Scarcelli G, Yun HS, Hahn SK, *Biomacromolecules* 13(8), 2554–2561, 2012.
299. Wang HJ, Zhang J, Liu YH, Luo TY, He X, Yu XQ, *RSC Adv.* 7, 15613, 2017.
300. Wang L, Yin Y, Jain A, Zhou HS, *Langmuir* 30(47), 14270–14275, 2014.
301. Yang YH, Cui JH, Zheng MT, Hu CF, Tan SZ, Yang YXQ, Liu YL, *Chem. Commun.* 48, 380, 2012.
302. Du F, Zhang M, Li X, Li J, Jiang X, Li Z, Hua Y, Shao G, Jin J, Shao Q, Zhou M, Gong A, *Nanotechnology* 25, 315702, 2014.
303. Li H, Kong W, Liu J, Yang M, Huang H, Liu Y, Kang Z, *J. Mater. Chem. B* 2, 5652, 2014.
304. Hsu P, Shih ZY, Lee CH, Chang HT, *Green Chem.* 14, 917–920, 2012.
305. Zheng M, Ruan S, Liu S, Sun T, Qu D, Zhao H, Xie Z, Gao H, Jing X, Sun Z, *ACS Nano* 9(11), 11455–11461, 2015.
306. Li H, Kang Z, Liu Y, Lee ST, *J. Mater. Chem.* 22, 24230–24053, 2012. doi:10.1039/C2JM34690G.
307. Datta KKR, Kozak O, Ranc V, Havrdova M, Bourlinos AB, Safarova K, Hola K, Tomankova K, Zoppellaro G, Otyepka M, Zboril R, *Chem. Commun.* 50(74), 10782–10785, 2014.
308. Lv C, Lin Y, Liu A-A, Hong Z-Y, Wen L, Zhang Z, Zhang Z-L, Wang H, Pang D-W, *Biomaterials* 106, 69–77, 2016.
309. Du Ting, Liang J, Dong N, Liu L, Han H, *Carbon* 110, 278–285, 2016.

310. Su XL, Li Y, *Anal. Chem.* 76, 4806, 2004.
311. Weng CI, Chang HT, Lin CH, Shen YW, Unnikrishnan B, Li YJ, Huang CC, *Biosens. Bioelectron.* 68, 1, 2015.
312. Ghosh M, Sonkar SK, Saxena M, Sarkar S, *Small* 7, 3170, 2011.
313. Dong Y, Li G, Zhou N, Wang R, Chi Y, Chen G, *Anal. Chem.* 84(19), 8378–8382, 2012.
314. Zhang YL, Wang L, Zhang HC, Liu Y, Wang HY, Kang ZH, Lee ST, *RSC Adv.* 3(11), 3733–3738, 2013.
315. Gao X, Lu Y, Zhang R, He S, Ju J, Liu M, Li L, Chen W, *J. Mater. Chem. C* 10(3), 2302, 2015.
316. Dong Y, Wang R, Li G, Chen C, Chi Y, Chen G, *Anal. Chem.* 84, 6220, 2012.
317. Zhu A, Qu Q, Shao X, Kong B, Tian Y, *Angew. Chem.Int. Ed.* 51, 7185 2012.
318. Liu X, Zhang N, Bing T, Shangguan D, *Anal. Chem.* 86(5), 2289–2296, 2014. doi:10.1021/ac404236y.
319. Zong J, Yang X, Trinchi A, Hardin S, Cole I, Zhu Y, Li C, Muster T, We G, Biosens. Bioelectron. 51, 330–335, 2014.
320. Gedda G, Lee CY, Lin Y-C, Wu H-F, *Sens. Actuators B Chem.* 224, 396–403, 2016.
321. Qu KG, Wang JS, Ren JS, Qu XG, *Chem. Eur. J.* 19(22), 7243–7249, 2013.
322. Zhang H, Chen Y, Liang M, Xu L, Qi S, Chen H, Chen X, *Anal. Chem.*86(19), 9846–9852, 2014.
323. Lu W, Qin X, Liu S, Chang G, Zhang Y, Luo Y, Asiri AM, Al Youbi AO, Sun X, *Anal. Chem.* 84(12), 5351–5357, 2012.
324. Liu HR, Li HT, Kong W, Liu J, Liu Y, Tong C, Zhang X, Kang Z, *Mater. Res. Bull.* 48(7), 2529–2534, 2013.
325. Zhang R, Chen W, *Biosens. Bioelectron.* 55, 83–90, 2014.
326. Seabra AB, Durán N, *J. Mater. Chem.*20, 1664, 2010.
327. Seabra AB, Durán N, *Curr. Nanosci.* 8, 520, 2012.
328. Seabra AB, Marcato PD, de Paula LB, Durán N, *J. Nano Res.*20, 61, 2012.
329. Seabra AB, de Lima R, Calderon M, *Curr. Top. Med. Chem.*15, 298, 2015.
330. Yu CM, Li XZ, Zeng F, Zheng FY, Wu SZ, *Chem. Commun.*49, 403–405, 2013.
331. Gao X, Ding C, Zhu A, Tian Y, *Anal. Chem.* 86, 7071, 2014.

332. Zhang LL, Han YJ, Zhu JB, Zhai YL, Dong SJ, *Anal. Chem.* 87, 2033–2036, 2015.
333. Ni P, Dai H, Li Z, Sun Y, Hu J, Jiang S, Wang Y, Li Z, *Talanta* 144, 258–262, 2015.
334. Bhaisare ML, Pandey S, Khan MS, Talib A,Wu HF, *Talanta* 132, 572–578, 2015.
335. Zhang X, Ming H, Liu R, Han X, Kang Z, Liu Y, Zhang Y, *Mater. Res. Bull.* 48, 790–794, 2013.
336. Wu HY, Mi CC, Huang HQ, Han BF, Li J, Xu SK, *J. Lumin.*132, 1603–1607, 2012.
337. Shi W, Li XH, Ma HM, *Angew. Chem. Int. Ed.* 51, 6432, 2012.
338. Jin XZ, Sun XB, Chen G, Ding LX, Li YH, Liu ZK, Wang ZJ, Pan W, Hu CH, Wang JP, *Carbon* 81, 388–395, 2014.
339. Ding C, Zhu A, Tian Y, *Acc. Chem. Res.* 47(1), 20–30, 2014.
340. Bai W, Zheng H, Long Y, Mao X, Gao M, Zhang L, *Anal. Sci.* 27(3), 243–246, 2011.
341. Noh EH, Ko HY, Lee CH, Jeong MS, Chang YW, Kim S, *J. Mater. Chem. B* 1, 4438- 4445, 2013.
342. Maiti S, Das K, Das PK, *Chem. Commun.* 49, 8851–8853, 2013.
343. Jana J, Ganguly M, Pal T, *Phys. Chem. Chem. Phys.* 17, 2394–2403, 2015.
344. Deng JH, Lu QJ, Hou YX, Liu M L, Li HT, Zhang YY, Yao SZ, *Anal. Chem.* 87, 2195–2203, 2015.
345. Li H, Chen L, Wu H, He H, Jin Y, *Langmuir* 30, 15016, 2014.
346. Li Q, Ohulchanskyy TY, Liu R, Koynov K, Wu D, Best A, Kumar R, Bonoiu A, Prasad PN, *J. Phys. Chem.C* 114(28), 12062–12068, 2010.
347. Zhai X, Zhang P, Liu C, Bai T, Li W, Dai L, Liu W, *Chem. Commun.*48, 7955–7957, 2012.
348. Du F., Min Y, Zeng F, Yu C, Wu S, *Small* 10, 964, 2014.
349. Martorana A, Koch G, *Front. Aging Neurosci.* 6, 252, 2014.
350. Matsumoto M, *Mov. Disord.* 30, 472, 2015.
351. Suzuki T, Kanahara N, Yamanaka H, Takase M, Kimura H, Watanabe H, Iyo M, *Psychiatry Res.* 227, 278, 2015.
352. Li H, Liu J, Guo S, Zhang Y, Huang Y, Liu Y, Kang Z, *J. Mater. Chem. B* 3, 2378, 2015.
353. Li Q, Xu Z, Tang W, Wu Y, *Anal. Lett.* 48, 2040, 2015.

354. Jiang Y, Wang B, Meng F, Cheng Y, Zhu C, *J. Colloid Interface Sci.* 452, 199, 2015.
355. Lin S, Reppert J, Hu Q, Hudson JS, Reid ML, Ratnikova TA, Rao AM, Luo H, Ke PC. *Small* 5(10), 1128–1132, 2009.
356. Jorio A, Braulio SA, Ribeiro-Soares J, Pagano MC, Ferreira EHM, Araujo JR, Silva AM, Capaz RB, Newton PSF, Cançado LG, Achete CA. Chap. 6 - Study of carbon nanostructures for soil fertility improvement, in: *Bioengineering Applications of Carbon Nanostructures, Nanomedicine and Nanotoxicology*, Jorio A (ed.), doi:10.1007/978-3-319-25907-9_6, Springer International Publishing Switzerland, 2016.
357. Sharon M, Sharon M, *Carbon Nano Forms and Applications*, McGraw Hill, USA, 2009.
358. Mauter MS, Elimelech M, *Environ. Sci. Technol.* 42, 5843–5859, 2008.
359. Liu X, Zheng M, Xiao Y, Yang Y, Yang L, Liu Y, Lei B, Dong H, Zhang H, Fu H, *ACS Appl. Mater. Interfaces* 5(11), 4667–4677, 2013.
360. Jackson P, Jacobsen NR, Baun A, Birkedal R, Kuhnel D, Jensen KA, Vogel U, Wallin H, *Chem. Cent. J.* 7(1), 154, 2013.
361. Li W, Zheng Y, Zhang H, Liu Z, Su W, Chen S, Liu Y, Zhuang J, Lei B. *ACS Appl. Mater. Interfaces* 8(31), 19939-19945, 2016.
362. Kerfahi D, Tripathi BM, Singh D, Kim H, Lee S, Lee J, Adams JM, *PLoS One* 10(2015), e0123042, 2015.
363. Khodakovskaya M, Dervishi E, Mahmood M, Xu Y, Li Z, Watanabe F, Biris AS, *ACS Nano* 3, 3221–3227, 2009.
364. Khodakovskaya M, Kim BS, Kim JN, Alimohammadi M, Dervishi E, Mustafa T, Cernigla CE, *Small* 9, 115–123, 2013.
365. Khodakovskaya M, Silva KD, Biris AS, Dervishi E, Villagarcia H, *ACS Nano* 6, 2128–2135, 2013.
366. Lu CM, Zhang CY, Wen JQ, Wu GR, Tao MX, *Soybean Sci.* 21, 68–172, 2002.
367. Cañas JE, Long M, Nations S, Vadan R, Dai L, Luo M, Ambikapathi R, Lee EH, Olszyk D, *Environ. Toxicol. Chem.* 27, 1922–1931, 2008.
368. Tripathi S, Sonkar SK, Sarkar S, *Nanoscale* 3, 1176–1181, 2011. doi:10.1039/C0NR00722F.

369. Larue C, Pinault M, Czarny B, Georgin D, Jaillard D, Bendiab N, Mayne-L'Hermite M, Dive MTV, Carrière M, *J. Hazard. Mater.* 227, 155–163, 2012.
370. Miralles P, Johnson E, Church TL, Harris AT, *J. Royal Soc. Interface* 9, 3514–3527, 2012.
371. Dillow C, Carbon Nanotubes Shown to Boost Plant Growth, Could Spawn Super-Fertilizers Science, *Popular Science*, Sept. 25, 2009.
372. Tripathi S, Sarkar S, *Appl Nanosci.* 5, 609–616, 2015.
373. Qu SN, Wang XY, Lu QP, Liu XY, Wang LJ, *Angew. Chem., Int. Ed.* 51, 12215–1221, 2012.
374. Berestovsky GN, Ternovsky VI, Kataev AA, *J. Exp. Biol. Bot.* 52(359), 1173–1177, 2001.
375. Lin D, Xing B, *Environ. Sci. Technol.* 42, 5580–5585, 2008.
376. Zheng Y, Xie G, Zhang X, Chen Z, Cai Y, Yu W, Liu H, Shan J, Li R, Liu Y, Lei B, *ACS Omega* 2(7), 3958–3965, 2017. doi:10.1021/acsomega.7b0065.
377. Monreal CM, McGill WB, Nyborg M, Can. *J. Soil Sci.* 66, 499–511, 1986.
378. De Rosa MC, Monreal C, Schnitzer M, Walsh R, Sultan Y, *Nat. Nanotechnol.* 5, 91, 2010.
379. Kottegoda N, Imalka M, Nadeesh M, Veranja K, *Curr. Sci.* 101(1), 73–76, 2011.
380. Yao Y, Gao B, Inyang M, Zimmerman AR, Cao X, Pullammanappallil P, Yang L, *J. Hazard. Mater.*190, 501–507, 2011.
381. Atkinson CJ, Fitzgerald JD, Hipps NA, *Plant Soil* 337, 1–18, 2010.
382. Qian ZS, Shan XY, Chai LJ, Ma JJ, Chen JR, Feng H, *Biosens. Bioelectron.* 60, 64–70, 2014.
383. Martens DC, Westermann DT, Fertilizer applications for correcting micronutrient deficiencies, in: *Micronutrient Deficiencies in Global Crop Production*, Alloway BJ (ed.), Springer, pp. 549–553, 1991.
384. Ashfaq M, Verma N, Khan S. *Environ. Sci. Nano* 4(1), 138–148, 2017. doi:10.1039/C6EN00385K.
385. Kumar R, Ashfaq M, Verma N, *J. Mater. Sci.* 53, 7150–7164, 2018.

386. Rai V, Acharya S, Dey N, *J. Biomater. Nanobiotechnol.* 3, 315–324, 2012.
387. Vedamalai M, Periazhao AP, Wang CW, Tseng YT, Ho LC, Shih CC, Chang HT, *Nanoscale* 6(21), 13119–13125, 2014.
388. Gómez-de PS, Salinas-Castillo A, Ariza-Avidad M, Lapresta-Fernández A, Sánchez-González C, Martínez-Cisneros CS, Puyol M, Capitan-Vallvey LF, Alonso-Chamarro JS, *Nanoscale* 6(11), 6018–6024, 2014. doi:10.1039/c4nr00573b.
389. Zheng M, Xie Z, Qu D, Li D, Du P, Jing X, Sun Z, *ACS Appl. Mater. Interfaces*5(24), 13242–13247, 2013. doi:10.1021/am4042355.
390. Lee HN, Ryu JS, Shin C, Chung HJ, *Macromol. Biosci.* 17(9), 1700086, 2017. doi: 10.1002/mabi.201700086.
391. Ding L, Yang H, Ge S, Yu J, *Spectrochim. Acta A Mol. Biomol. Spectrosc.* 193, 305–309, 2018.
392. Lu S, Wu D, Li G, Lv Z, Chen Z, Chen L, Chen G, Xia, You J, Wu Y, *RSC Adv.* 6, 103169–103177, 2016.
393. Bandi R, Reddy B, Dadigala GR, Eslavath R, Singh SS, Guttena V, *RSC Adv.* 6, 28633, 2016.
394. Ju J, Chen W, *Biosens. Bioelectron.* 58, 219–225, 2014.
395. Koman VB, Tedrick TS, Lew MH, Wong H, Kwak SY, Giraldo JP, Strano MS, *Lab Chip* 17, 4015–4024, 2017. doi:https://doi.org/10.1039/c7lc00930e
396. Raeyani D, Shojaei S, Kandjani SA, Wlodarski W, *Procedia Eng.* 168, 1312–1316, 2016.
397. Rad F, Mohsenifar A, Tabatabaei M, Safarnejad MR, Shahryari F, Safarpour H, Foroutan A, Mardi M, Davoudi D, Fotokian M, *J. Plant Pathol.* 94, 525–534, 2012.
398. Safarpour H, Safarnejad MR, Tabatabaei M, Mohsenifar A, Rad F, Basirat M, Shahryari F, Hasanzadeh F, *Can. J. Plant Pathol.* 34, 507–515, 2012.
399. Zheng Z, Zhou Y, Li X, Liu S, Tang Z, *Biosens. Bioelectron.* 26, 3081–3085, 2011.
400. Chouhan R, Aaydha V, Thakur MS. *Anal. Bioanal. Chem.* 397(4), 1467–1475, 2010. doi:10.1007/s00216-009-3433-1.
401. Ge S, Lu J, Ge L, Yan N, Yu J, *Spectrochim. Acta A Mol. Biomol. Spectrosc.* 79, 1704–1709, 2011.

402. Mezian MJ, Dong X, Zhu L, Jones LP, LeCroy GE, Yang F, Wang S, Wang P, Zhao Y, Yang L, Tripp RA, Sun YP, *ACS Appl. Mater. Interfaces* 8, 10761–10766, 2016.
403. Da Silva LC, Oliva MA, Azevedo AA, De Araujo MJ, *Water Air Soil Pollut.* 175, 241–256, 2006.
404. Eichert T, Kurtz A, Steiner U, Goldbach HE, *Physiol. Plant.* 134, 151–160, 2008.

Index

Absorbance, 37, 51–52, 117
Acacia Arabica, 25
Acidic oxidation, 17–18, 161
Adhesive, 102
Aerosol, 148
AFM, 34–35
Allotropes, 159
Amorphous, 16, 30, 33, 40, 149
Ampicillin, 115–116
Amphiphilic, 67, 182–183
Analytes, 228
Anchoring, 140, 194
Angiogenesis, 144–145
Annihilation of fluorescence, 96
Antibacterial, 104, 142
Antibiotics, 77, 98, 110–112, 114–116, 132, 141–142, 156, 191, 234, 237–240
Antibodies, 43, 119, 134, 154, 166
Anticancer drug, 77, 86, 91, 128–131, 138–140, 142, 145–146, 149, 178
Apoptosis, 68, 85, 108, 154, 176
Aqueous dispersibility, 12, 65, 229
Arc discharge, 10–11, 20–21
Ascorbic acid, 15–16, 62, 199, 206
ATCC 25922 cells, 61
Autoclave, 31

Bacteria, 98, 104, 111, 114–116, 163, 181–183, 213, 224–225, 238–240
Band gap, 59
Biochar, 211, 225
Biocompatibility, 9, 31, 43, 48, 68–70, 74–75, 82, 100, 132, 136, 146–148, 151, 155, 163, 173–174, 177, 221, 237
Biocompatible, 29, 31, 60, 69, 79, 81, 84–86, 88, 94, 99, 105, 112–113, 124, 129, 141, 171, 194, 212, 216
Biogenic, 11, 28, 76–78, 89, 93, 132, 136, 212
Bioimaging, 38, 48, 60–62, 64–65, 69–70, 74, 80–81, 85, 88, 92, 100, 103, 109, 136, 139, 146, 159, 161–163, 166, 168, 170, 172, 175–176, 178, 180–183, 194, 206, 213
Biosensing, 76, 80, 89–91, 100, 184
Biosensors, 227
Black suspension, 20
Blinking nature, 64
Blood brain barrier, 69, 74, 118, 126, 137, 150, 164–165
Blood serum, 203
Blue shift, 7, 121
Blue-photoluminescence, 85, 98, 105
Bohr's exciton radius, 7, 48–49
Boron, 99–100, 226
Bovine serum albumin, 17. 79. 86, 129, 138, 141, 168, 178
Breast cancer, 79, 99, 108, 143, 149, 170, 176, 242
Breast tumor cells, 170
Bright luminescence, 23, 59, 79, 89

Calcination, 27
Cancer, 17, 22, 60, 78–79, 84–88, 91, 99, 101, 105, 108–110, 124, 128–131, 134, 138–140, 143–144, 147–149, 154, 157, 163, 166, 168, 170, 173–174, 176, 178, 191, 204, 242, 254–255
Cancer therapy, 85, 105, 128, 144, 148
Cantilever, 34
Carbon nanofiber, 17
Carbon nanomaterials, 213
Carbon nanotubes (CNT), 2, 9, 17, 72–73, 109–110, 213–214, 261
Carbonization, 14–15, 22–23, 26–28, 49, 52, 89, 175, 185, 203
Carboplatin, 128
Catalysts, 9
CD conjugates, 74, 80, 111, 119, 128–129, 140, 145, 147
CdS, 95–96
Cell membrane, 79, 86, 151–152, 162, 168, 170, 174–178, 182, 216–218, 236, 240
Cell wall, 115–116, 210, 216–220, 232–235
Cellular imaging, 89, 107, 129
Chemical approaches, 26
Chemical inertness, 48, 67, 69–70, 132
Chemical methods, 10–11, 26
Chemical moieties, 52
Chemical oxidation, 10
Chemical reactions, 137
Chemical stability, 109, 144, 206, 228
Chitosan, 18, 64, 75, 88–89, 151, 166, 175, 182, 226
Ciprofloxacin, 39–40, 111–114, 142
Cis-platin, 128, 130–131
Colorectal cancer, 131, 176
Combustion, 11, 13, 44, 161
Composite, 14, 27–28, 71–73, 75, 89–90, 92–93, 95–98, 110, 141–142, 151, 203, 227
Composition, 22, 31, 33, 41, 65, 87, 153, 221, 228
Condensation, 26, 30, 78, 103, 151
Condensation reactions, 98, 131
Conducting ink, 230
Conductivity, 47, 197–198, 218
Conjugate, 17, 40, 52, 55–56, 71–82, 84–92, 94–112, 114–119, 121–125, 127–132, 134, 137–142, 145–147, 149, 154, 156, 164, 168–174, 179, 181, 224, 226, 237–238, 240, 254
Covalent bonds, 72, 77, 81
Curcumin, 125–127, 157
Current density, 11, 137
Cuticular, 210, 233–235, 237
Cyclodextrin, 105–106, 114, 142, 156
Cysteamine hydrochloride, 77, 106, 120–123, 150
Cytoplasm, 79, 83–84, 128, 130, 140, 146, 149, 162, 170, 173–176, 178–179
Cytotoxic, 68, 100, 103, 108, 137, 160, 176
Cytotoxicity, 68–69, 75, 82, 91, 97, 100, 103, 109, 130–131, 136–137, 140, 149, 160, 162–163, 165, 169–170, 172, 175, 184, 188, 194, 206, 212

Dangling bonds, 110
D-band, 44
De Broglie wave length, 2, 6
Delocalized electrons, 72
Dendrimer, 77, 87, 102, 107
Diagnostic, 3, 50, 84, 127, 132, 181
Diffusion, 22, 149–150, 163, 220, 235–237
Digitonin, 89
Dihydro-lipoic acid, 106
DNA, 76, 78, 80–83, 91, 103–105, 112, 129–130, 140–141, 143, 145, 151, 181, 184, 200–202
Dopamine, 190, 204, 206
Dopant, 99, 136
Doxorubicin, 81, 86, 92, 125, 128–131, 139, 146, 148, 178–179

Drosophila, 78, 183
Drug delivery, 17, 25, 38, 48, 69–70, 75–78, 80–81, 89, 91–92, 99, 107, 109, 111, 119–121, 126, 128–142, 144, 146–150, 152–153, 155–157, 165, 211, 237
Drug loading, 74, 92, 112, 117, 124, 130, 137, 147, 156, 238
Drug release, 74, 76–77, 88, 92, 113, 117–118, 120, 124, 134, 137–140, 146, 149–150, 156
Dual Polarization Interferometry (DPI), 42–43, 45

EDX, 89, 215
Ehrlich Ascite Carcinoma, 173
Electric field, 13
Electrochemical exfoliation, 203
Electrochemical method, 11–13, 22, 61, 91–92, 137, 161
Electro-chemi-luminescence (ECL), 24
Electron diffraction, 33
Electron transport, 75
Electronic devices, 159
Electronic properties, 2–3, 48, 50, 100
Electronic transitions, 7, 51, 60
Encapsulated, 74, 81, 89, 104, 106, 156, 223
Encapsulation, 75, 77
Energy storage, 224
Enzyme, 81, 110, 116, 146, 174, 196, 202–203, 232
Erwinia117, 239
Ethylene diamine, 15, 175
Europium, 76, 97–98

Fe Amino Clay (FeAC), 109
FE-SEM, 108
Fertilizer, 210–212, 214, 219, 221–228, 240, 261
Filtration, 17
Fire blight of apple, 239
Fluorescence, 3, 9–10, 14, 16, 18–19, 23–25, 29, 31, 37–38, 45, 47, 53–69, 75, 78–84, 88–90, 94, 96–100, 103, 106–108, 127–129, 131, 138–139, 141, 146–147, 151–152, 154, 161–170, 172–177, 179–180, 182–198, 200–202, 206, 216, 228–229, 231–232
Fluorescent carbon, 9, 159
Fluorescent material, 54
Fluorescent sensor, 87, 185, 191, 194, 201
Fluorophore, 57, 60, 94
Food packaging, 210
Foliar, 117, 224, 226, 233–234, 236–237, 239–240
Folic acid, 17, 76, 86–87, 92, 129, 138, 154, 168, 178
Fourier Transform Infra-Red (FTIR), 9, 16, 20, 22, 24, 38–41, 44–45, 121, 123, 126, 238
Fullerenes, 211
Functional groups, 21, 25, 30, 38, 41, 44, 52, 65, 67, 99, 107, 154–155, 161, 185, 189–190, 204
Functionalization, 12, 14, 19, 21, 48, 52, 67–68, 70, 74–76, 81, 110, 121, 128, 136–137, 144, 160–161, 166, 173, 199, 204, 228

Gadolinium (III), 97
Gastric cancer, 84–85, 174, 176
Gelatin, 16
Gel electrophoresis, 13, 20
Gene therapy, 82, 101, 151, 232
Glioblastoma, 173
Glucose, 16, 18, 22, 25–26, 28, 52, 103–105, 161, 169, 175, 177, 202–203
Glutathione, 131, 166, 229
Glycerol, 25, 62, 103, 140, 146, 161, 199
Glycol, 18, 23, 25, 57, 64, 80, 96, 101, 103, 135, 161, 185, 192, 199

Graphene, 2, 9–10, 12, 17, 20, 40, 43, 72, 79, 109, 135, 159, 161, 166, 212, 226, 231, 242
Graphene oxide, 58, 178, 192, 203
Graphene quantum dots(GQD), 10, 20, 79, 135, 159–160, 229, 231
Graphite, 10–13, 17–19, 43, 59, 72, 109, 136, 161, 197–198, 203
Graphitic, 11, 18, 41, 147, 159
Green beans, 215
Gum Arabic, 29–30, 39, 112, 117, 237

Haloperidol, 106, 119–121, 123, 150
Healthcare, 181, 246
Heat treatment, 185
HeLa cells, 81, 86, 103, 109, 129, 140, 146, 168–169, 178, 194, 200, 204
Helium, 21
Hepato cellular carcinoma, 175–176
Heterocarbon, 10, 43, 90, 99–100, 135, 159, 161, 190
Hg, 100, 140, 185, 190–193
H_2S, 194
Hole, 3–7, 49, 53, 56–57, 59, 66, 95–96, 184, 218–219, 228
HOMO, 62
HRTEM, 33, 113
Hydrocarbon, 15, 182, 234
Hydrogen peroxide, 16, 22, 41, 110
Hydrophilic, 13–14, 120, 132, 171
Hydrophobic, 21, 29–30, 77, 107
Hydroquinone, 195–196
Hydrothermal oxidation, 14–16
Hydrothermal method, 16–18, 25, 89, 127, 151, 163–164, 173, 182, 192, 200
Hydroxyl, 16, 22, 25, 30, 38, 40, 67, 72, 81, 98–99, 123, 132, 147, 161, 197, 200, 238

Imaging, 3, 32–33, 60, 69, 78–79, 81–82, 84–85, 87, 89–92, 97, 100–101, 103, 107, 109, 113, 129–130, 138, 147, 150, 152, 154, 162–163, 165–168, 170–171, 173–174, 176–178, 180–181, 184, 187–188, 191, 200, 203
Immobilization, 43, 203
Intrinsically, 63, 100, 114, 160, 228

Lack of Blinking, 63, 70
Laser Ablation, 10–11, 18–19, 59, 161, 170, 191
Lattice, 110
Lipid, 76–78, 86, 97, 119, 150, 181, 220, 234
Living tissues, 200
Luminescence, 16, 23, 50, 53–54, 58–60, 63, 79, 89, 94, 100, 145, 164–165, 170, 185, 188, 192, 195, 202, 227, 256
LUMO, 62
Lung cancer, 85, 91, 105, 148, 176

Magnetic, 6, 44, 138, 162
Mammalian, 78–79, 101, 151, 168, 177
MCF-7, 68, 79, 99, 108, 149, 154, 162, 170, 188
MDCK cells, 29
Mechanical, 109–110, 126, 218, 230
Mechanical properties, 135
Micronutrient, 226, 228, 261
Microscopic methods, 32
Microwave, 23–25, 30, 51, 84, 120, 161, 174, 182
Microwave assisted, 11, 23–25, 29, 52, 64, 80, 103, 112, 120, 122, 125, 140, 150, 168, 170, 188, 195, 206, 215, 237
Microwave irradiation, 23, 84, 97
Middle lamella, 217–218
Morphology, 22, 33, 75, 128, 173, 177, 211, 221
Multi-photon excitation, 47, 60–61, 70
Multi wall carbon nano tubes (MWCNT), 77, 109–110, 160, 213

Nano-fertilizer, 210–212, 221–224
Nano-lithography, 10
Nano-medicine, 139, 255–257, 260
Nanoscale, 73, 98, 109, 153, 160, 212, 221, 223, 227, 245, 247–249, 252, 254, 256, 260, 262
Nanotechnology, 47, 148–150, 152, 155, 209–211, 221, 223–225, 227–228, 231–232, 257
Negative/ negatively, 80, 82, 85, 102, 104, 112–116, 126, 128, 130, 139–140, 142, 151–152, 155, 168, 174, 212, 220, 228, 239
Neurodegenerative, 77, 106, 118–119, 126, 134, 150, 204
Neurons, 120, 125–126, 204
Nickel oxide (NiO), 20
NIR, 22, 26, 50, 60–62, 88, 92, 147–148, 151, 165, 167, 184
NIR light excitation, 148, 166–167, 184
Nitric acid, 13–14, 17–19, 21, 44, 59, 68, 173
Nitrogen, 9, 65, 81, 84, 87, 99–100, 103, 127, 135, 159, 163, 173, 183, 186, 190, 192, 195, 216, 221–225
Nitrogen doped, 24, 62, 116, 127, 149, 163–164, 186
Nutrient, 144, 211–213, 216–222, 224–225, 227–228, 234, 240
NMR, 44–45, 126, 172
Nucleic acid, 83, 200–201, 229
Nucleus, 1, 79, 84, 99, 101, 128–130, 140, 144, 146, 149, 174–175, 178–179

Oligonucleotides, 151
One-dimensional (1D), 2, 159
Open-circuit voltage, 75
Optical absorption, 51, 56, 160
Optical properties, 7, 9, 37, 39, 49–50, 70, 83, 131, 146, 160–162, 228, 238
Optical sensors, 188, 192
Optoelectronic device, 108, 132
Organophilic, 14
Oxidants, 66
Oxidizing agents, 29, 52
Oxaliplatin, 131, 139

PAMAM, 87, 107
Paraffin, 13
Particle, 2–6, 9–10, 12, 23, 28, 30–31, 41–42, 49–50, 55–56, 59, 74, 89, 92, 97, 118–119, 136, 146, 163, 168, 173, 184, 211, 220–221, 227, 236–237
Patents, 153, 209
Payload, 75–76, 82, 86, 92, 113, 129, 136, 211, 213, 216, 219, 237, 240
PCR, 174
PEG, 19, 28, 68–69, 75, 80, 85, 89, 96, 100–101, 108, 114, 146, 151, 163–164, 166, 171, 181–182, 192, 204
Permeability, 91, 119, 157, 188, 234–235
Peptide, 76–79, 86, 101, 119, 129, 134, 180
PET (Photoinduced electron transfer), 52, 102, 106
Phosphate, 11–12, 81–82, 97–98, 173, 193–194, 224
Phosphorus, 25
Photobleaching, 23, 48, 63–65, 70, 132, 147
Photocatalytic property, 65, 70, 95
Photodynamic therapy, 147
Photo-excited, 52–53, 65, 148, 184
Photogenerated, 95–96
Photoluminescence, 9–10, 12, 19, 22, 47, 50, 53–54, 60–62, 65, 70, 82, 84–85, 98, 104–105, 127, 174, 185, 190, 198, 205–207, 228
Photostable, 79, 91
Physical methods, 11
Phytic acid, 25
Planck-Einstein Equation, 6
Plant-derived, 25, 28

Plasma, 11–12, 81,142, 153–154, 161, 163, 206, 218, 233, 235–236
Plasma membrane, 218, 233, 235–236
Plasma treatment, 11–12, 161
Plasmodesma, 218
Plasmon, 94
PLGA, 75, 166
PMAA, 210
Polar solvents, 48, 148
Polarity, 16
Pollutant, 132
Poly-ethyleneimine (PEI), 18, 68–69, 82–85, 88, 101–105, 140–141, 146, 151–152, 165–166, 169, 172, 174, 187–188, 204
Polymer, 43, 45, 67, 72–73, 75–77, 82–83, 88, 98, 101–102, 135, 137, 140, 156–157, 166, 204, 222, 226, 233–234
Polysaccharide, 25, 88, 105, 155, 217, 218
Pore size, 150, 230
Potassium, 201, 225–226
Primary cell wall, 217–218
Properties of carbon dots, 45, 47
Protein, 17–18, 45, 69, 76–79, 83, 85–86, 91, 101, 106–107, 116, 119, 124–126, 134, 141, 145, 154, 163, 172–173, 176, 181–182, 184, 201–202, 204, 220–221, 236, 238
Pulsed laser irradiation, 20
Pyrolysis, 11, 14, 24, 80, 103, 140, 146, 195

Quantities, 2, 116, 213, 223, 239
Quantum confinement, 2–4, 7, 10, 26, 48–49, 52, 55–56, 231
Quantum dots (QD), 2–10, 20, 26, 31, 40, 43, 45, 47–48, 59–60, 63, 68–69, 79, 94, 103, 119, 132, 135, 159–160, 162, 164, 180–181, 185, 187, 191, 206, 209, 212, 216, 228–229, 231–232, 254
Quantum mechanics, 2, 4, 6, 45
Quenching, 16, 87, 98, 127, 182, 184–185, 187–190, 192, 196, 228–229, 231

Raman scattering, 91
Raman spectroscopy, 29, 43, 91, 211
Reactive oxygen species (ROS), 108
Renal cancer, 147–148
Resistance, 9, 63, 69–70, 91, 98, 112, 148, 176, 240
Resonance, 44, 108, 127, 147, 162, 192, 212, 232
Reverse micelles, 26, 28
Rhodamine B, 95–96, 108, 188, 199
Rice husk, 93, 146
RNA, 78, 80–81, 83, 152, 174, 182

Schizophrenia, 119, 150, 204
Secondary cell wall, 218
SEM, 32, 45, 108
Semiconducting, 9, 59, 109, 162
Semiconductor quantum dots, 9, 31, 45, 60, 63, 68, 94, 181, 184, 228
Sensing, 73–74, 80–81, 88, 103, 106, 108–109, 115, 127, 174, 185–189, 193, 195, 197, 201, 204, 206, 226–228
Sensitive, 74, 90, 98, 100–101, 103, 130, 138, 184, 186, 188, 194, 198–200, 204, 229, 232, 235
Sensor, 16, 38, 43, 87–88, 99–100, 108, 110, 161, 184–185, 187–199, 201–204, 206, 227, 229–233
Silica, 27–28, 76–77, 82, 91–94, 109, 146, 188, 232
SiRNA, 82–85, 105, 152, 174
Solvothermal method, 18, 103, 175, 186
Sorbitol, 17, 24, 44, 86, 129, 138, 168, 178
Spatial dimension, 4, 8
Spleen, 163
STM, 34, 35
Streptomycin, 116–118, 237–240
Stoke's shift, 55

Stomata/stomatal, 219, 230, 233–234, 236–237, 239
Strontium oxide, 96
Sucrose, 14, 16, 25, 29, 161, 166
Sulphur, 75, 99–100
Surface area, 74, 118, 153, 210, 219, 222–223
Surface emissive traps, 52
Surface passivation, 12, 19, 23, 25–26, 28, 31, 38, 45, 52, 55–58, 64–65, 67–68, 75, 82, 160, 164, 166, 184
Surface state, 57, 216
Surfactant, 27–28, 92, 196, 234, 237
SWCNT, 9, 17, 20, 109, 161

TEM, 32–33, 45, 84, 89, 220
Template, 14, 26, 28
TNT (2,3,6-trinitro toluene), 195
Tetracycline, 98, 114, 141–142, 156, 240
TGA, 29, 238
Thermal oxidation13–15, 28, 97, 136, 168
Thermal treatment, 14, 28
Thrombin, 81–82, 202
TiO_2, 66, 138
Titanium, 13
Toxic/ toxicity, 3, 48, 68–70, 74, 91–92, 98, 100–102, 128, 132, 135, 140–142, 144, 147–148, 150–151. 157, 160, 164, 173, 175–176, 183–184, 191, 212, 227, 232, 236
Transduction, 78
Tunable, 14, 47, 54, 65, 75, 103, 147, 160, 162, 173, 177, 191, 199, 212
Transferrin, 119, 124–125, 168–169, 204
Trichomes, 233–234, 237
Two-dimensional (2D), 2, 8, 159, 234

Ultrasonication, 22, 25
Unstable, 57
Upconversion photoluminescence, 16, 47, 54, 61–62, 66, 70, 89, 94, 165, 188, 198
Urea, 24, 115, 127, 215–217, 222–223
UV-Vis Spectroscopy, 37–38, 238

Vancomycin, 114–115
Vehicles, 48, 80–81, 86, 92, 107, 111–112, 120, 126, 129, 136, 149–150, 152–154, 157, 237
Visible irradiation, 65
Vitamins, 184, 201
Vitamin C, 221

Water solubility, 9, 16, 67, 70, 74, 107, 132, 170, 203, 221, 228
Water treatment, 102
Water soluble, 22, 75, 84, 138, 173, 182, 189, 224
Wheat, 214–215
Work function, 102

X-Ray Photoelectron Spectroscopy/ XPS, 41–42, 44, 216
X-Ray Diffraction/ XRD, 29, 40, 45

Zero-dimensional (0D), 8, 135, 159
ZnO, 59, 66, 76, 94–95

Also of Interest

Check out the other titles in the "Advances in Nanotechnology & Applications" series

Behaviors and Persistence of Nanomaterials in Biomedical Applications
By Domenico Cassano and Valerio Voliani
Published. ISBN 978-1-119-41827-6

Carbon Dots as Theranostic Agents
By Madhuri Sharon and Ashmi Mewada
Published. ISBN 978-1-119-46010-7

The History of Nanotechnology: From Prehistoric to Modern Times
Edited by Madhuri Sharon
Forthcoming October 2018. ISBN 978-1-119-46008-4

Fundamentals of Electrocatalyst Materials and Interfacial Characterization: Energy Producing Devices and Environmental Protection
By Nicolas Alonso-Vante
Forthcoming October 2018. ISBN 978-1-119-46007-7

WWW.scrivenerpublishing.com

Lightning Source UK Ltd.
Milton Keynes UK
UKHW021329260319
339922UK00003B/70/P

9 781119 460107